Mathematical Go

Mathematical Go
Chilling Gets the Last Point

Elwyn Berlekamp
U C Berkeley
Berkeley, California

David Wolfe
U C Berkeley
Berkeley, California

A K Peters
Wellesley, Massachusetts

Editorial, Sales, and Customer Service Office

A K Peters, Ltd.
289 Linden Street
Wellesley, MA 02181

Library of Congress Cataloging-in-Publication Data

Berlekamp, Elwyn R.
    Mathematical go: chilling gets the last point / Elwyn Berlekamp, David Wolfe.
      p. cm.
    Includes bibliographical references and index.
    ISBN 1-56881-032-6
    1. Go (Game) 2. Go (Game) — Mathematical Models. I. Wolfe, David. II. Title.
GV1459.55.B47   1994
794—dc20   93-40859
                                    CIP

Printed in the United States of America
99 98 97   10 9 8 7 6 5 4

Contents

Foreword

This is an extraordinary book. The senior author (Elwyn Berlekamp) plays Go at only the 10-kyu level, and his colleague David Wolfe is rated an amateur shodan, yet they have developed techniques to solve late-stage endgame problems that stump top-ranking professional players. The problems typically offer a bewildering choice of similar-looking moves, each worth only one or two points, but with subtle priority relationships that cannot be adequately described by sente and gote. The solutions come out of combinatorial game theory, a branch of mathematics that Berlekamp helped develop. A Go player who masters its techniques can extract a one-point win from positions where the uninitiated will almost invariably lose or draw.

The theory presented in this book assigns each active area on the board an abstract value, then shows how to compare these values to select the optimum move, or add them up to determine the optimum outcome. Some of the values are familiar numbers or fractions, but most are more bizarre objects (arrows, stars, trees) quite unlike anything in the existing Go literature. From these abstractions, the reader will learn that positions seeming to have the same numerical value can be crucially different, while positions that appear completely different can be mathematically identical.

It should be emphasized that this book will not help the reader improve his opening or midgame strategy. Nor does it reveal any tactical secrets; locally, the tactics are all of the simple kind that a beginner can read out easily. In terms of practical benefit on the board, the most the reader can hope for is to get stronger by one point. (Of course, a lot of games are decided by one point.)

A Go player with an interest in mathematics, however, or a mathematician interested in Go, will relish this book, because it describes substantial connections between the two subjects which have hitherto been largely unrecognized. The theory developed is precise, rigorous, intellectually appealing, and demonstrably successful. As a bonus, there is a novel discussion of the mathematical rules of Go.

September 1993

James Davies
Tokyo, Japan

Preface

History of combinatorial game theory and its role in Go

As a typical game of Go approaches its conclusion, the active regions of play become independent of one another. Play in each region is not affected by play in the others; although one player may make several successive plays in the same region while his opponent chooses to play elsewhere. The overall board position therefore may be regarded as a *sum* of disjoint partial board positions.

Combinatorial game theory has long been concerned with such sums of games. The first nontrivial result was perhaps the solution of the game of Nim by Bouton at Harvard in 1901. In the 1930s, his work was extended to a broader class of games by Sprague and Grundy; later, Guy and Smith [GS56] [Guy89] masterfully exploited this theory to obtain complete, closed-form solutions to a wide variety of games. Most of their work dealt only with *impartial* games, in which all pieces on the board have the same color and can be moved by either player. These games have no scores; rather, the outcome depends only on who got the last move. Nonetheless, the impartial theory later played a major role in the analysis of Dots and Boxes [BCG82, pp. 507–550], even though it is a contest over scores.

In the early 1970s, Conway discovered and axiomatized the theory of combinatorial games, including *partisan* games, which generalized impartial games so that players might control pieces or stones of opposite colors. Conway also discovered several theorems about the special role of games called *numbers*, including the *mean value theorem*, which provided a new formulation of earlier work by Hanner [Han59] and Milnor [Mil53] on sums of games with scores. Many results were published in *On Numbers and Games* [Con76]; more results plus numerous examples appeared in *Winning Ways* [BCG82], a two-volume treatise on which Berlekamp, Conway, and Guy had collaborated since the late 1960s.

These results made it possible to analyze games in which score determined the winner, and so it was tempting to apply the theory to Go. Initial attempts to apply combinatorial game theory to Go endgames in the 1970s failed, partly because certain crucial notions (particularly, *chilling* and *warming* operators) were not yet sufficiently well understood.

By the early 1980s, combinatorial game theory had developed powerful techniques for handling many other games in which board positions tend to break up into disjoint active regions. These included Hackenbush, Toads and Frogs, and Domineering. Here, the theory assigns an abstract *value* to each active region. In a simple inactive region, this value coincides with a numerical count of the score. In an active region, this value embodies the traditional Go player's notion of the *count*, as well as a considerable amount of additional information about the local situation. Each value depends only on a local analysis of the relevant partial board position. The value does *not* depend on who moves next; it takes all competent sequences of local play into account. And fortunately, the values of most simple positions all belong to a relatively small set of common simple values. Each value can occur in many different positions. So, to analyze the entire board position, one can compute the mathematical *sum* of all of the mathematical values representing the various regions of the board.

However, the practical utility of such a theory depends very much on the properties of the values which occur. If the local positions are sufficiently complicated that their values are intractable, then the sum can be very hard to analyze. Until the late 1980s, it had appeared as though the values which occur in even simple Go positions were so strange and intractable as to limit the theory's usefulness.

In 1984, Robin Pemantle [Pem84] calculated values for many $2 \times n$ and $3 \times n$ Domineering positions. Efforts to understand the properties of these values led to refined *warming* and *overheating* operators, and to a new game called Blockbusting, which can be viewed as a simplified variation of $2 \times n$ and $3 \times n$ Domineering. Chilled Blockbusting values are well-understood numbers, and, in Blockbusting, an appropriate warming operator inverts chilling.

This book

In 1989, it became clear that *chilling* reduces many common Go positions to familiar values, such as numbers, infinitesimals, and switches, and that chilling can be inverted by an appropriate warming operator. The authors of this book met and our collaboration began. Much of this book is based on Wolfe's dissertation [Wol91].

After chilling, many values which are common in late-stage Go endgames turn out to be very tractable, even though these values and their mathematical properties are not yet well known outside a small community of specialists. The values which appear are universal; they apply directly to Winning Ways games and to the ancient Hawaiian game of Konane. They facilitate a complete, precise, and thoroughly rigorous analysis of sums of Go endgames.

Acknowledgements

Since 1990, significant contributions to Mathematical Go have been made by a number of students from four incarnations of Berlekamp's combinatorial game theory courses, including Raymond Chen, David Moews, David Moulton, and Yonghoan Kim.

Our work on this book has also benefitted from the advice, criticism, and novel ideas of many others. We received support from the Massachusetts Institute of Technology, from the Center for Pure and Applied Mathematics of the University of California at Berkeley, and from Mitsubishi Electric Research Laboratory, where Berlekamp was a Visiting Fellow in the spring of 1992. During the past 3 years, both of us have given a number of formal and informal seminars and lectures on this subject. We have received some very constructive ideas and comments from many persons, including Eric Baum, David Kent, Kiga Yasuo, Martin Mueller, Jurg Nievergelt, Nick Patterson, Lisa Stewart, Herb Taylor, and Takizawa Takenobu. Stephen Parrott, James Davies, and Richard Bozulich reviewed early versions of this book and suggested so many major improvements that it was substantially rewritten. Richard Bozulich has also served as the facilitator, coordinator, and translator for our visits to Japan, and Fujisawa Kazunari has coordinated our interactions with the professional Go community at the Nihon Ki'in.

We have grown increasingly appreciative of the very different perspectives from which games are viewed by Go players, mathematicians, and computer scientists. People who routinely visualize the subject from more than one angle are rare, but they do exist. One such person was Bob High, who was President of the American Go Association at the time of his untimely death in a white water rafting accident on January 8, 1993. He was a strong amateur Go player who also had a keen interest in mathematics. He attended several of our seminars and learned a great deal about this subject by studying the papers and notes that we sent him. He then wrote two papers, [Hig91] and [Boz92, pp. 218–224], [Boz92, pp. 218–224], aimed at making some version of this material accessible to the Go player who had no training in advanced mathematics. These manuscripts played a pivotal role in persuading us of the feasibility of this book.

We believe the study of mathematical Go is still in its infancy. Generalizations of the notions of *temperature* and *mean values* to include many types of Kos are already known. Progress is continuing rapidly, and we are hopeful that others will refine and extend the results presented in this book.

List of Figures

Chapter 1

Introduction

Since you are reading this book, you are likely to fall into one of three categories:

- A mathematician interested in the applications of mathematics to games and Go.[1]

- A Go player interested in how mathematics might improve your game.

- A computer scientist interested in how to design or improve Go playing programs.

Mathematicians will find theorems about Go both subtle and surprising. Combinatorial game theory provides general methods for analyzing games, particularly games in which typical positions have separate pieces with limited interaction. This decomposition into pieces gives *group* structure, and this structure can be exploited to precisely analyze Go endgame positions that don't yield to the methods of the strongest Go players.

Go players can find quicker ways to improve their game than to read this book. Many of the positions analyzed do not tend to come up in contexts which take full advantage of the subtleties of the results. However, there are some wonderful lessons for the Go player. Sente (the worth of keeping the initiative), and gote (giving up initiative) are issues which are only vaguely understood by all but the finest Go players. This is in part because the

---

[1] A top-down overview of the rules to Go appears in Appendix A. In this book, Japanese or American style scoring rules are assumed (i.e., scoring territory minus prisoners). Extensions of the results to rules which score territory plus stones on the board are discussed in Appendices A and B. Also, a glossary of terms appears in Appendix F.

concepts are subtle, but also because the meaning of initiative is, by nature, vague and amorphous. The game theory does away with these concepts by providing clearer and more concrete methods, and these in turn will give the Go player a better understanding of sente and gote. Lastly, the technique of "playing the difference game" is basic to combinatorial game theory, and does not seem to exist in traditional Go theory; the Go player is encouraged to concentrate much of his/her attention on this technique given in Section 3.3.

For the computer scientist, the mathematical theory provides general and precise methods for simplifying local game trees in the endgame. These methods take into account the possibility that a player may move several times in a row in a local position when the opponent chooses to move elsewhere. Unfortunately, a virtually complete description of local game trees is required; the methods are therefore most applicable at a point when the typical game can be analyzed by brute force. Nonetheless, these methods are a start in a promising new direction. A good program will require a strong understanding of how to integrate what is known about nearly separated positions.

1.1   Why study Go

People want to understand the things that people like to do, and people like to play games. Perhaps that's all the justification needed to study games. But games also provide a concrete, self-contained framework in which to study mathematical and programming methods. In a game not only are the rules (and therefore the model) clear, but one measure of success is clear: How does a human or computer play relative to an experienced player with established records or ratings?

Go is particularly attractive to study because:

- Go's three or four thousand year old history and popularity throughout the world (particularly in Asia) means there are many experts on whom theories can be tested and from whom insights can be gained.

- Simple rules shorten the process of designing models, and make the game susceptible to mathematical analysis.

- Go poses new and more formidable challenges for the sorts of programming methods which have had great success in chess. Go is therefore an excellent testing ground for new artificial intelligence techniques.

In fact, over the past ten years, Go has become an increasingly popular game for studying tree search, pattern recognition, learning, planning, and knowledge bases. For information about computer Go, a list of over a hundred references can be found in [Kie90]. Also, a quarterly journal called *Computer Go* [Edi] contains current articles and computer tournament results.

In short, Go is a popular game with enticing problems for both the mathematician and the computer scientist.

1.2   Easy (?) endgame problem

Belle Black and Wright White sit down to continue a game where part of the board is temporarily obscured. Belle has been assured that there are no kos on the rest of the board, and that the best move is at either *a* or *b* in Figure 1.1. Further she knows that all stones are alive with the exception of the five stones labeled ⬤. Which move is better for Belle Black — *a* or *b*? Or does it depend on what is on the rest of the board? Or perhaps it makes no difference, no matter what is on the rest of the board?

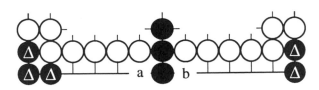

Figure 1.1: *Which move is better for Black?*

If you're like most strong Go players, you will think that surely *a* is at least as good as *b*. After all, not only does the move at *a* pose a larger threat (to save three stones rather than two), but the threat comes sooner (after three moves rather than four). So either it makes no difference which move is made (they're each worth a point), or perhaps *a* is better in some unusual positions.

Well, this analysis turns out to be wrong — *b* is sometimes the better move, and *a* is never better! Given this, you may think *b* is better because there are more one-point moves available on that side. In fact, this latter explanation is also flawed, for in Figure 1.2, move *a* is the better move.

So, how can you possibly be convinced that move *b* is better for Black in the first position (Figure 1.1)? One start might be to see an example where

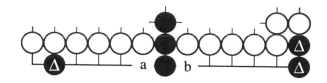

Figure 1.2: *Reducing the number of stones on the left makes a the better move*

b is, indeed, better. Consider the position on the left in Figure 1.3 where White has just moved at ①. Assume that the obscured middle of the board has been completely resolved to a tie in score. So whoever does better in the last four endgame regions will win the game.

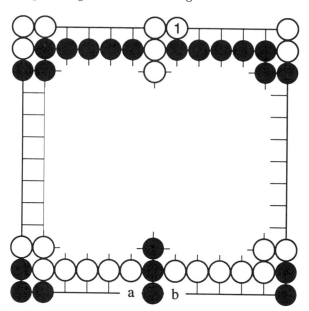

Figure 1.3: *A position where b is better*

If Belle Black responds at *b*, surely she can get a tie. The bottom of the board mirrors the top, and no matter what Wright White does from then on, Ms. Black simply maintains symmetry. Suppose, on the other hand, Black responds at *a*. Then, as we'll see in the diagrams that follow, Mr. White can proceed to win by one point, and *b* is indeed the better move!

One possible line of play is shown in Figure 1.4(a). Black simply blocks White's incursions, and then White blocks Black's. In this line, White gets the last move and consequently wins by one point. In fact, **getting the last move is the key** — when Black played the symmetric play at *b* in Figure 1.3, she could get a tie by maintaining symmetry throughout the game to guarantee the last play. It seems that Black's plays at ❷ and ❹

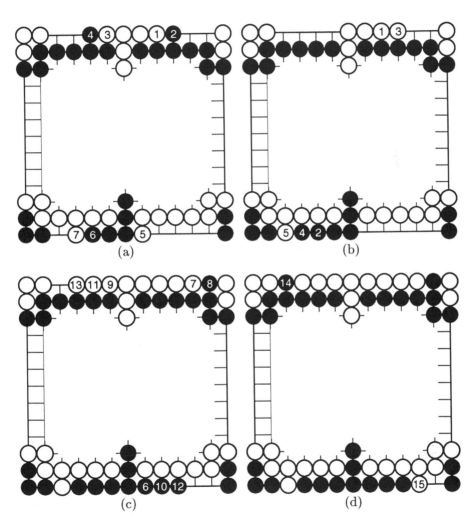

Figure 1.4: *Two possible lines of play*

are *gote* (i.e., gave up the initiative), using up any chances to get the last move.

This suggests a different sequence of play for Black (Figure 1.4(b–d)). Rather than simply blocking White's incursions, Black makes incursions of her own, blocking only when necessary. But in the end, the result is the same. Black would like to counter ⑬ at ⑮, but alas White's threat to save three stones is larger than Black's threat to save two.

These last two lines of play are not the only possibilities, but it is possible to prove that White can always win by one point. White's strategy is to continue his incursions, the upper right one taking priority, and only block Black's incursions when necessary. The reader is invited to try to explore Black's other lines of attack.

So we have argued that *b* is better than *a* in *some* positions (again, Figure 1.1). But maybe there are positions when *a* is better? Well, the search for such a position might begin in a similar fashion as Figure 1.3 earlier. In particular, suppose White moves at ① in Figure 1.5, and Black responds at ❷. In this case, Black can get a tie, and one natural line of play is shown. Notice that Black need not use ⑯ to block White's incursion, since she can make the bigger threat to save three stones.

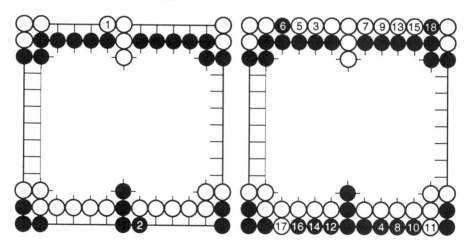

Figure 1.5: *What if White plays the other way?*

It's possible to mathematically prove that the move at *b* is better than the move at *a* in Figure 1.1 no matter what else is on the board, as long as the opponent has no opportunities to create a ko. Further, demonstrating that *a* leads to a loss in Figure 1.3 and that Figure 1.5 leads to a tie is sufficient for such a proof! This is the method of using difference games to compare moves, which will appear in more detail in Section 3.3.

1.3 Teaser

After reading this book, the industrious reader will be rewarded by being able to solve problems such as those in Appendix C, one of which appears in Figure 1.6. The object is to play all of the small endgame moves in just the right order to get the last point. This problem has stumped several 9-dan professionals from Japan and China, and no one has solved it without the techniques from combinatorial game theory. The diligent reader who masters the relevant mathematics will understand the solutions in Appendix D.

If you are a strong amateur Go player, and believe this problem looks easy, you are not alone. Unfortunately, there is little that can be said briefly to convince you otherwise. One property of this problem that differs from most in the Go literature (e.g., [Nak84]) is that there is no one key move; all of the first twenty or more moves are equally difficult, and on each of these moves very few choices lead to a winning position. So, even if a Go player thinks he has the answer, verifying it requires a human or computer opponent well versed in the theory.

1.4 Useful programs

Raymond Chen has written a package of PC software [Che] which uses the mathematical analysis to play all of the problems collected in Appendix C. That software package is available from Ishi Press [Ish]. Those readers who wish to test their understanding of this book will enjoy playing those problems against the software. The machine is willing to play either Black or White. If you let it play White, you have no chance; if you let it play Black, it should still be a worthy opponent.

A computer program written in C for UNIX implements many of the combinatorial game theory concepts [Wol]. It converts expressions to canonical form, recognizes many of the usual games, cools, computes atomic weights, etc. The program is valuable in analyzing games, generating and testing hypotheses, or just experimenting with the theory.

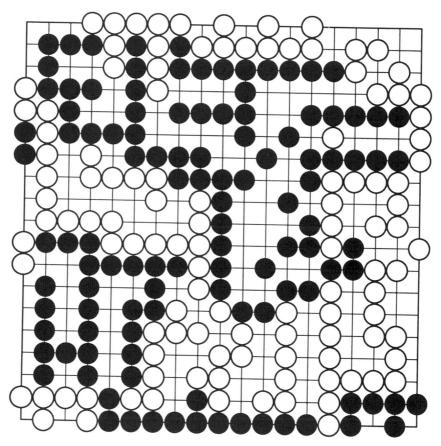

Figure 1.6: *White to move and win*

Chapter 2

An Overview

By the end of this section, you will be able to solve problems that even many top professionals cannot solve! (Problems of varying difficulty appear in Appendix C.) This section serves as a cookbook for small endgame positions. Simple guiding rules will be introduced to solve endgame positions, progressing quickly from simple positions to rather complicated ones. However, why the rules work will not be addressed until later in the book.

2.1 Fractions

Consider the endgame problem in Figure 2.1. For each lettered area (a through f), how much is the position worth to Black (or White) as it stands? And how much is the move worth to Black? Or to White?

To figure this out, refer to Figure 2.2 which lists *blocked corridors*.[1] The number of points of territory Black has is listed in the second column, and how much a move is worth (to either Black or White) is listed in the third column. One mnemonic for remembering the table is that the integer value for the number of points each region is worth (4 for the last corridor) is the number of points Black keeps when White plays first and Black responds as

---

[1]The values shown can be made precise if the overheating symbol $\int$ (not yet defined) is used. For instance, the value of the last corridor should be $4 \int \frac{1}{32}$ and the move is worth $\int \frac{31}{32}$. The reasons will become evident later on in the book. Also, *temperature* is not yet defined. For now, it is sufficient to think of temperature as a numeric estimate on the value of a move. The units are half the gote-value of a move in Japanese Go literature.

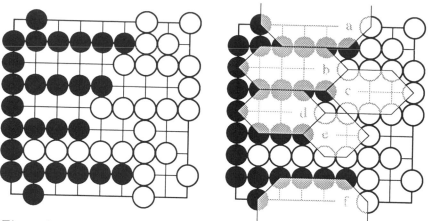

Figure 2.1: *White to move and win — How much is each move worth?*

below:

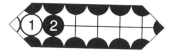

The fractional part is given by $\frac{1}{2^n}$, where n is the number of points Belle Black gets when she moves first. This would be $\frac{1}{2^5} = \frac{1}{32}$ for the last corridor:

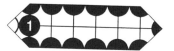

Using this method, it is possible to evaluate the number of points available to both players in each unplayed region as in Figure 2.3. On the left is the number of points as calculated above — positive values refer to Black points, and negative values refer to White points. Since we'll want to focus on the fractions rather than the integer part of each region, we'll use the shorthand shown in the righthand Figure 2.3. A black (respectively, white) dot or *marking* is placed on the board for each point of Black (respectively, White) territory. The same mnemonic mentioned in the previous paragraph suggests the way the dots are placed here. For each region, assume the attacker moves first, and place a dot on each point of territory remaining. The fractions refer only to the adjustments required to match the numbers in the left diagram.

It is all well and good to pretend the territory and moves have fractional values, but what do they all mean? After all, when the game ends, one

POSITION	$\approx$ AREA	TEMPERATURE
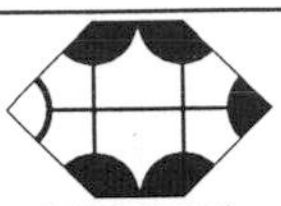	$\frac{1}{2}$	$\frac{1}{2}$
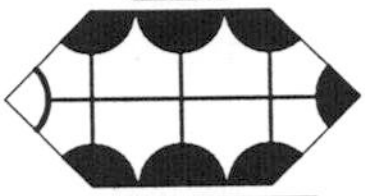	$1\frac{1}{4}$	$\frac{3}{4}$
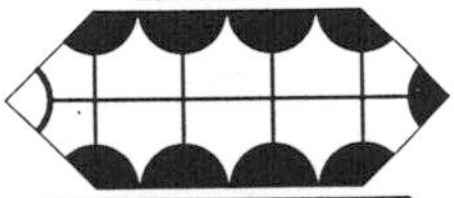	$2\frac{1}{8}$	$\frac{7}{8}$
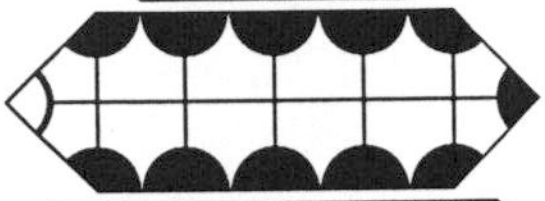	$3\frac{1}{16}$	$\frac{15}{16}$
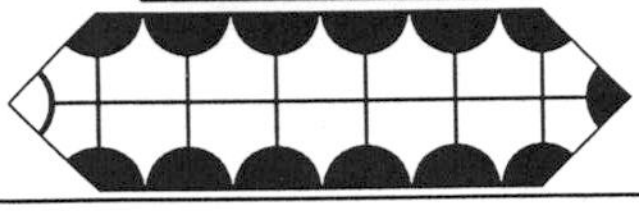	$4\frac{1}{32}$	$\frac{31}{32}$

Figure 2.2: *Each corridor with its value for Black, and the value of a move (Partially shown stones are safe.)*

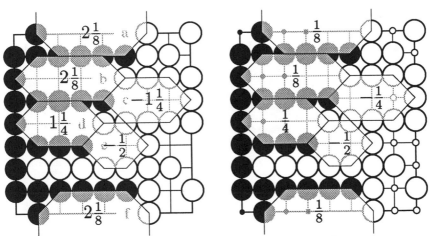

Figure 2.3: *The number of points each region is worth*

player will have some integral number of points rather than, say, $5\frac{1}{4}$ points. But the $\frac{1}{4}$ does have meaning! In particular, to find out how the score will end up at the end of a perfectly played game, just add up these fractional points for each player; the player on move will be able to round the fraction in his favor. For instance, if the sum were $5\frac{1}{4}$, and it is Black's move, Black should win by 6 points. If it is White's move, Black should win by 5 points.

In the example in Figure 2.3, there are an equal number of black and white dots, and the fractions add up to $-\frac{1}{8}$. Recall that a negative number is good for White. Therefore, if it is Wright's move he should win by 1 point, while if it is Belle's move the game should end in a tie. To win, Wright must play on one of the long corridors. Moves a, b, and f are each worth $\frac{7}{8}$ points, leaving Wright in the lead by exactly 1 point since $-\frac{1}{8} - \frac{7}{8} = -1$. If White were to move at c or d, he would only gain $\frac{3}{4}$ points, and the total would now be $-\frac{1}{8} - \frac{3}{4} = -\frac{7}{8}$. Since it's now Belle's turn, she'll round the fraction in her favor yielding a tie score. The move at e is worse still, and a, b and f are the only winning moves.

The corridors in Figure 2.2 were all *blocked* — the invader was invading from one end of the corridor. It turns out that *unblocked corridors* (intrusions from both ends) are also easy to evaluate. Their values are shown in Figure 2.4, along with their markings (dots).[2]

You may be concerned that the temperature is less than you'd predict. For instance, Belle's move from the position $\approx 2\frac{1}{8}$ to the position $\approx 3\frac{1}{16}$ gains about $\frac{15}{16}$ rather than the $\frac{7}{8}$ in the temperature column.[3] The mathematics will bear this out, but a Go player's less precise explanation may alleviate your concerns for now. Since the moves at the ends of unblocked corridors are *miai*, we can assume, when evaluating a position and its temperature, that Black will take one end and White will take the other, and so = .

---

[2]The reader may be bothered by the white markings in the topmost corridors in Figure 2.4. These markings are convenient so that the pattern of adding one black dot for each added node in corridor length is maintained. Consequently, many theorems to come will be greatly simplified.

[3] When a stone is played at the end of a corridor such as , sometimes we'll depict the corridor by and sometimes . The meaning of both are the same, and the choice depends on what emphasis is desired in the text.

BLOCKED	UNBLOCKED	UNMARKED UNCHILLED		MARKED CHILLED	
		$\approx$ AREA	TEMP.	AREA	INC.
		$-2 + 2$	0	2	
		$-1 + 1$	$1 - 1$	1	
	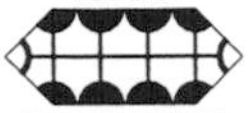	$0 + \frac{1}{2}$	$1 - \frac{1}{2}$	$\frac{1}{2}$	$-\frac{1}{2}$
	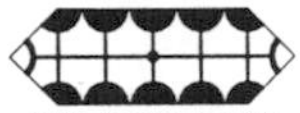	$1 + \frac{1}{4}$	$1 - \frac{1}{4}$	$\frac{1}{4}$	$-\frac{1}{4}$
	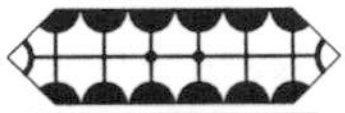	$2 + \frac{1}{8}$	$1 - \frac{1}{8}$	$\frac{1}{8}$	$-\frac{1}{8}$
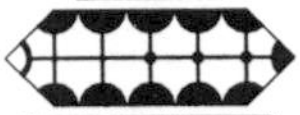	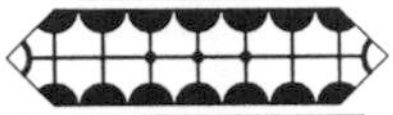	$3 + \frac{1}{16}$	$1 - \frac{1}{16}$	$\frac{1}{16}$	$-\frac{1}{16}$
	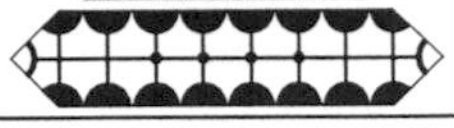	$4 + \frac{1}{32}$	$1 - \frac{1}{32}$	$\frac{1}{32}$	$-\frac{1}{32}$

Figure 2.4: *Values of corridor areas, unchilled temperatures and chilled incentives*

2.2 Chilling

For long corridors, the value of a move is rather close to one point. Since most good moves in these positions gain around a point, we measure the value of a move with respect to that point. Rather than saying the value of the second to last corridor in Figure 2.4 is $\frac{15}{16}$, we'll say the move gains $-\frac{1}{16}$ (or loses $\frac{1}{16}$). We'll refer to $-\frac{1}{16}$ as the *chilled* value as a reminder that one point has been taxed away. Later, when analyzing more interesting Go positions, this methodology will not merely be convenient, but tremendously useful, simplifying many complicated equations. For now, please accept the methodology even though its advantages may not yet be fully appreciated.[4]

In line with this methodology, each of Belle's moves typically adds a black marking (or removes a white marking) and each of Wright's adds a white marking. For example, from ⟨▬▬▬▬▬⟩, Black moves to ⟨▬▬▬▬▬⟩

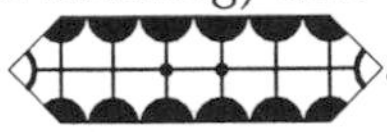

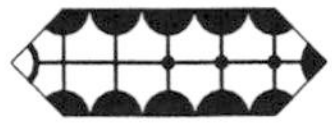

---

[4]The reader familiar with mathematical games will recognize G chilled as G_1, G cooled by 1. The $\int$ symbol is used to invert chilling and will be discussed more later.

and White moves to ◇──────◇. Enforcing that the player who moves must add a marking of his color is called *playing the chilled game*. The adjustment of one marking accounts for the one-point tax on moving.

2.3 The need for more than just numbers

As most Go players already know, knowing how many points each move is worth is not enough to find the best move. Sente and gote, for instance, are issues. And there are times when which of two moves is correct depends on a part of the board completely unrelated to the two moves.

For example, consider the two problems in Figure 2.5. There are few enough possible moves that it's easy to work through all move sequences. The possibilities for the left position are further reduced by observing a and c are miai. On the left the only winning move is at b, and on the right the only winning move is at a. Yet the only difference between the positions is the available move at c. The various lines of play are shown in Figure 2.6. In short, whether a or b is correct depends on whether move c is available. It seems that it is impossible to assign a value to each move which always predicts the best play for even simple endgame problems.

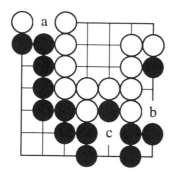 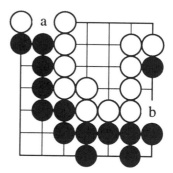

Figure 2.5: *White to move and win. (Both positions are identical except for the region around c.*

Combinatorial game theory provides a key to understanding these different sorts of positions. It provides new values (other than numbers) for these positions, and techniques for analyzing the properties of the values. All the new values are *infinitesimals* — smaller than all positive numbers and greater than all negative numbers. Just as fractional values are use-

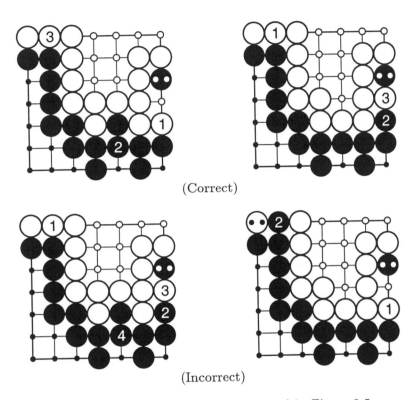

(Correct)

(Incorrect)

Figure 2.6: *Solutions to problems posed in Figure 2.5*

ful even though they don't appear in the final score, infinitesimals are also tremendously useful for evaluating a position and finding the best play.

2.4   Ups, downs and stars

Positions b, d and f in the problem in Figure 2.7 are already understood; they have values $\frac{1}{2}$, $-\frac{1}{4}$, and $-\frac{1}{4}$ respectively. To analyze the others, we'll need to understand a few more values.

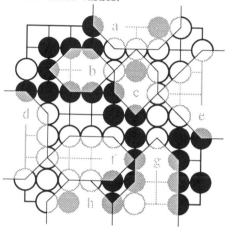

Figure 2.7: *White to move and win — How much is each move worth?*

The following position is a simple 2-point gote for both players:

If Belle moves she captures a single stone for two points, and Wright's move leaves zero. The following notation is used to denote the Black and White moves: within the braces, Black's move(s) are listed to the left of the bar, and White's move(s) are listed to the right.

$$\text{⬤⬤◯╫◯} = \left\{ \text{⬤⬤◯⬤◯} \;\middle|\; \text{⬤◯◯◯} \right\} = \{2 \mid 0\}$$

(Although Black's move captures one stone, the captive has been left on the diagram.)

Now let's consider the same position marked and chilled. (Clipping off part of the stones in the diagram indicates we're playing the chilled game.) Since in the original position, the average outcome is one point, it makes sense to use a single black marking:

Recall that when playing the chilled game, each move *must* add a marking:

$$\text{⬡} = \left\{ \text{⬡} \,\middle|\, \text{⬡} \right\} = \{0 \mid 0\}$$

When Black moves, she gets two points in the unchilled game, but the two points get taxed away by the markings leaving 0. A White move leaves no markings, and is also to 0. We call the game $\{0 \mid 0\}$ *star* and write it simply "$*$".

Notice that a simple unchilled dame is also worth $*$, for Belle and Wright can both move to 0:

$$\text{⬡} = \text{✚} = *$$

Again, remember that the position on the left is chilled while the one on the right is not.

It's worth noting that a chilled dame is worth 0:

$$\text{◇} = 0$$

This is because the tax on moving on the chilled game is so great as to discourage either player from doing so.

As a consequence,

$$\text{⬡} = \left\{ \text{⬡} \,\middle|\, \text{⬡} \right\} = \{1* \mid -1*\}$$

$$\text{} = \left\{ \text{} \;\middle|\; \text{} \right\} = \{0\,|\,0\} = *$$

(When chilling, the $*$'s disappear from $\{1*\,|-1*\}$.)

So how much is $*$ worth compared with fractions like $\frac{1}{32}$? It turns out that $*$ is *infinitesimal*, which means it's less than all positive numbers and greater than all negative numbers. And how do we do arithmetic with $*$? Well, it's very different from conventional numbers. First of all, $* + * = 0$ and $* = -*$ (even though $* \neq 0$), so it's most unlike any number. Further, $*$ is incomparable with 0, which means it's neither less than nor greater than nor equal to zero. So $*$ is a different sort of beast altogether.

Since $* + * = 0$, the possible values introduced so far are numbers (such as $5\frac{1}{4}$), perhaps plus $*$. Typically we write $n + *$ for numbers n simply as $n*$. So possible values include -2, 0, $3\frac{3}{8}$, $*$, $1*$, $2*$ and $-3\frac{1}{8}*$. As mentioned earlier, if Black is winning by some fractional number of points (such as $5\frac{1}{4}$), under perfect play the person on move will be able to round the fraction in his direction. So if Belle moves first, she should win by 6 points; if Wright first, 5 points. What if the value of the position totals to $5*$ or $5\frac{1}{4}*$? Well, $*$ is less than all positive numbers and greater than all negative numbers. So, $5\frac{1}{4}*$ gets rounded just like $5\frac{1}{4}$. But, $5*$ behaves differently: If Belle moves first, she wins by 6 points, if Wright moves first Black wins by only 4 points. The $*$ gets rounded to 1 or -1 depending on who moves first. This is a consequence of being an infinitesimal incomparable with 0. (Naturally, if the value were simply 5, Black would win by 5 points under optimal play, no matter who moves first.)

So now we understand the values of c and h in the last figure (Figure 2.7). Both regions have chilled value $-1*$. We're left with a, e and g.

The following chilled position has a new value discovered by John Conway, who christened it *up*, $\uparrow = \{0|*\}$:

$$\text{} = \left\{ \text{} \;\middle|\; \text{} \right\} = \{0|*\} = \uparrow$$

Again, since we are playing the chilled game, each move (by either player) adds a marking. We have no hard and fast rule for how to place the markings; they are placed for convenience to normalize the game to somewhere near 0.

Like $*$, $\uparrow$ is also infinitesimal: It's less than all positive numbers and greater than all negative numbers. Unlike $*$, $\uparrow$ is positive. In fact, $\uparrow + \uparrow$ is yet another infinitesimal called *double-up* and written $\Uparrow$. We can also have

⇑↑, ⇑↑↑, 5•↑, 6•↑, and so forth. Lastly, *down* is the negative of up (↓ = −↑ — the negative of a position is achieved by just reversing black and white). And we can write ⇓, ⇓↓, etc.

How do these games compare with *? Well, ↑ is incomparable with * and ⇑ > *. This is enough to know the final scores of games which have pieces like *, ↑ and numbers. Since ↑ + * is incomparable with 0, a game of value 3↑* should end with Belle winning by 4 or 2 points depending on who moves first. (Once again, we've left out the + symbol; 3↑* = 3 + ↑ + *.) But, 3↑ and 3⇑* both end with Belle winning by 4 or 3 points, as ↑ and ⇑* both exceed 0. Since all of these new games are infinitesimal, a game of value $3\frac{1}{4}$↑* will end in the same score as one of value $3\frac{1}{4}$; Belle wins by 4 or 3 points.

We wanted to know the values of positions *a*, *e* and *g* in the problem (Figure 2.7). Figure 2.8 gives a summary.[5] Notice that each position adds ↑* to the previous. Unlike numbers and *, the players have differing *incentives* to play; on a long corridor, Wright is much more eager to play since he gains ↑* by moving, while Belle's move typically loses all the ↑'s stored up in the position since her move leaves the position with value 0. This is akin to the Go proverb that a move in sente should be made before a gote move, but the mathematics give a more precise description. Notice also that White's eagerness to play on ⇑↑* equals his eagerness to play on ⇑↑, ⇑* or ↑, gaining ↑* by each play. But as soon as the game reaches *, White's eagerness drops as he gains only * by the play.

We now have enough machinery to crack that problem in Figure 2.7. Figure 2.9 summarizes the values of all the regions. The values of the regions can be summed:[6]

$$\underbrace{\frac{1}{2} + -\frac{1}{4} + -\frac{1}{4}}_{0} + \underbrace{↓ + ↓ + ⇑*}_{*} + \underbrace{* + *}_{0} = *$$

Wright can win the game by rounding the * in his favor. When playing the chilled game, White gains one marking. If the resulting total of chilled values is less than or equal to 0, Black can at best round the remaining game

---

[5] Note that the sign of White's incentive is chosen so that moves with larger incentive are good for White. This may appear inconsistent with the philosophy that White is the negative player. This choice of sign is convenient so that Black's and White's incentives can be compared.

[6] The regions *a*, *c* and *e* of Figure 2.9 do not appear completely independent. If Black gets to play in all the regions, the white stones in the upper right corner come under attack. This does not affect the values shown; the reasons will be discussed on pages 61–64.

POSITION	AREA	INCENTIVES BLACK	INCENTIVES WHITE
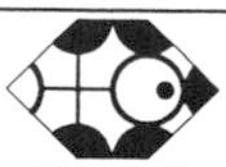	*	*	*
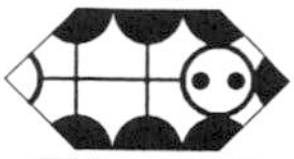	↑	↓	↑*
	⇑*	⇓*	↑*
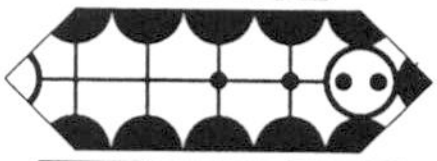	⇑⇑⇑	⇓⇓⇓	↑*
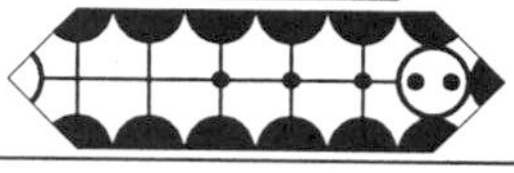	⇑⇑⇑*	⇓⇓⇓*	↑*

Figure 2.8: *Each corridor with its value for Black, and the value of a Black move and of a White move when playing the chilled game*

to 0; White will win by the one point accounted for in the marking. A move on a region of value * (at c or h) moves the new chilled total to 0 with one marking in his favor. The move on ⇑* (at g) moves the total to ↓. Any of these moves win. All other moves lose. If White moves on ↓ (a or e), the total would then be ↑* and Black could gain a tie by rounding in her favor. Any move on a number loses way too much. For instance, the move on $\frac{1}{4}$ would leave the total value of the game at $\frac{1}{4}*$ which is positive, and therefore Belle can gain back the marking.

The *best* play is the move on ⇑* in the sense that it gains the most, picking up ↑*. So, a player who has found all of the incentives but has not totalled up the value of the game can nevertheless guarantee that if any winning moves exist, the move at g will work.

Let's go back to the two problems introduced in Section 2.3. Figure 2.10 shows the value of each region. Figure 2.10(a) sums to ↓, so White should win. Either move on * moves to ↓* which is incomparable with 0 and therefore fails to win. The move on ↓ leaves a total value of 0, and with the gained marking gives a win. In Figure 2.10(b), the move on ↓ leaves * which is incomparable with 0 and fails to win. The move on * leaving ↓ wins.

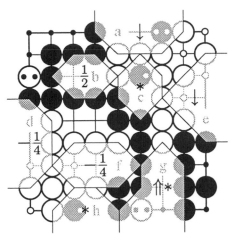

Figure 2.9: *The value of each region. Black and White have the same number of markings*

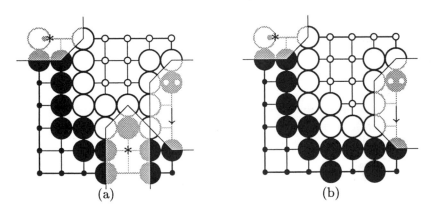

Figure 2.10: *Mathematical solution to Figure 2.5*

2.5 Tinies and minies

The following position looks like ↑, but Wright's move threatens to save more than one stone:

$$[\text{position}] = \left\{ [\text{position}] \mid [\text{position}] \right\}$$

$$= \left\{ [\text{position}] \parallel [\text{position}] \mid [\text{position}] \right\}$$

$$= \left\{ \quad 0 \quad \parallel \quad 0 \quad \mid \quad -2 \quad \right\}$$

Recall that $[\text{position}] = -2$ since the markings credit Black with two points which she does not have. The way to read a game equation with multiple |'s is to start at the place with the most |'s — in this case, ∥. On each side of the ∥ is a list of games (usually a list with just one game). Each game in the list is either a position (e.g., to the left of the ∥) or the possible moves from a position separated by fewer |'s (e.g., to the right of the ∥).

Conway christened this game $+_2$ (pronounced "tiny 2"); the subscript reflects the -2. If, after one move, White is threating to make an x point gote move (rather than the 4 points for the two stones here), we write $+_{x-2}$. So, $[\text{position}] = +_1$, $[\text{position}] = +_4$, and $[\text{position}] = \uparrow = +_0$ (though we'd much rather write ↑ than $+_0$, since ↑ has very different properties from your typical $+$). The negative of $+_2$ is $-_2$ (pronounced "miny 2"), and (like all negatives) is simply $+_2$ with the black and white stones reversed. $[\text{position}] = -_2$.

How do tinies compare with the other games like numbers, ↑ and *? Let x be any positive number (or, for that matter, any game exceeding a positive number — like $x = \frac{1}{2}$ or $x = \frac{1}{2}*$). Then $+_x$ is a positive infinitesimal that's less than ↑. In fact, $+_x$ is much much less than ↑; no matter how many $+$'s you add together, you'll never add up enough to get bigger than ↑. You could say that $+_x$ is infinitesimal with respect to ↑, just as ↑ is infinitesimal with respect to positive numbers.

This is all consistent with a Go player's instincts. If a move is sente for White, typically White will get to play it before Black does. And Black will be forced to respond and consequently $[\text{position}]$ is very nearly

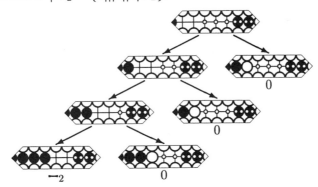 = 0. But Black has a slight preference for the former position, since in some cases she'll get to move to 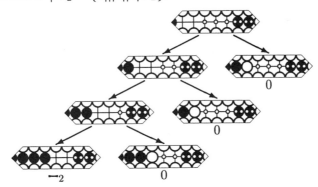 before White plays.

In fact, $+_x$ is even infinitesimal with respect to $+_y$ if $x > y$. For example, 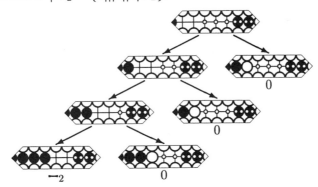 = $+_4$ is infinitesimal with respect to 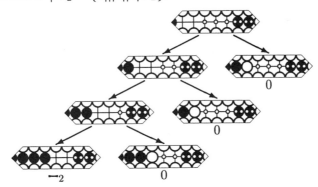 = $+_3$. An intuitive explanation is that a White move on $+_4$ requires a more urgent response than a move on $+_3$, and therefore $+_4$ is closer to 0 than $+_3$.

What happens when $+$'s with different subscripts get added together? Unfortunately the resulting games get rather complicated. Instead of worrying about what the values of the games are, we'll concentrate on where the best moves are when there are both $+$'s and other games around.

But before discussing the best moves, let's return to the sorts of positions introduced in Section 1.2. These were longer corridors ending in $+$'s and $-$'s like 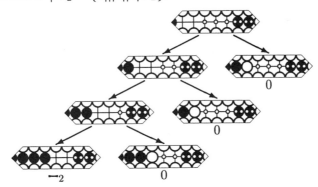. Figure 2.11 gives the game tree for this position, which we can summarize as:

$$\text{} = \{-_2|0\|0\|\|0\}$$

Each group of |'s indicates a position in the game tree which has followers: the more |'s, the higher up in the game tree. Each game value (0 or $-_2$ here) indicates a position which we already understand (at the bottom of the game tree). We'll abbreviate the game $\{-_2 \mid 0\|0 \mid\|\mid 0\}$ by $-_2 \mid 0^3$, meaning White has three chances to move to 0 before the game reaches $-_2$. The negative of $-_2 \mid 0^3$ is written $0^3 \mid +_2 = \{0\|\|\|0\|0\|+_2\}$.

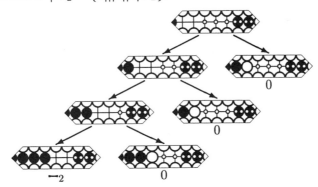

Figure 2.11: *The possible move sequences from $\{-_2 \mid 0\|0 \mid\|\mid 0\}$*

Examples (see pp. 189–192 for others)		
	$-y$	First plays should be attacks on minies and big gote moves. Better to attack a tiny, creating a $y + 2$ point gote move than to cash in on a $y + 2$ point gote.
or $\left\{\begin{array}{c} \text{} \\[2mm] \text{} \end{array}\right.$	or $\left\{\begin{array}{c} \{y \mid 0\} \\[4mm] \{0 \mid -y\} \end{array}\right.$	
	$-x$	Attacks on minies whose threats are smallest are worth least.
or $\left\{\begin{array}{c} \text{} \\[2mm] \text{} \end{array}\right.$	or $\left\{\begin{array}{c} \{x \mid 0\} \\[4mm] \{0 \mid -x\} \end{array}\right.$	
	$-x \mid 0$	Next, attack long corridors ending in bigger than 2 points gote moves. (Shorter corridors are attacked first if the gote moves at the end are equal.)
	$-x \mid 0^2$	
	$-x \mid 0^3$	
$\vdots$	$\vdots$	
	$-y \mid 0$	If two longer corridors are around, **attack the corridor with fewer stones at the end first!** (Approach the two stone group before the three stone group.)
	$-y \mid 0^2$	
	$-y \mid 0^3$	
$\vdots$	$\vdots$	
Continued on following page		

Figure 2.12: *(Text on page 26) Which corridors Black should attack first. Here, x and y are numbers with $y > x > 0$. In the examples, $x = 2$ and $y = 4$. Remember to ignore any pair of chilled games summing to zero. Positions higher in the table are more urgent for Black.*

Continued from prior page		
Examples (see pp. 189–192 for others)		
or $\left\{\vphantom{\begin{array}{c}a\\b\\\vdots\end{array}}\right.$	or $\left\{\begin{array}{c}\downarrow\\[4pt]\Downarrow*\\[4pt]\vdots\end{array}\right.$	Next attack corridors ending in one stone. Each such move is worth the same amount.
or ?	*(maybe)	But the move which connects the stone is worth less! Play a move on * if an odd number of them are around.
or $\left\{\vphantom{\begin{array}{c}a\\b\end{array}}\right.$	$\uparrow$ or $\Uparrow*$ or ... or $0^n|\!\!+_x$	Block *any* corridor ending in at least a two point gote.
$\vdots$	$\vdots$	
or $\left\{\vphantom{\begin{array}{c}a\\b\end{array}}\right.$	or $\left\{\begin{array}{c}\frac{1}{4}\\[6pt]-\frac{1}{4}\end{array}\right.$	Numbers are low priority. Play the number with larger denominator first.
or $\left\{\vphantom{\begin{array}{c}a\\b\end{array}}\right.$	or $\left\{\begin{array}{c}\frac{1}{2}\\[6pt]-\frac{1}{2}\end{array}\right.$	
	0	Dame are played last.

Just as with the $-$'s and $+$'s, these games get very complicated when they get added together. However, the following method always finds a winning move if one exists when the positions remaining on the board are those discussed so far. Assume you are playing Black, and refer to Figure 2.12.

1. First, pair off infinitesimals which are negatives of one another and therefore add to 0. Ignore these games. (A Go player will recognize this as ignoring positions which are miai.)

2. From the remaining games, the first priority is to attack long corridors and/or defend attacks on $+$'s. The order in which corridors are chosen to attack can be important, and is summarized in Figure 2.12.[7] Perhaps the most surprising fact is that corridors ending in *smaller* threats should be attacked first.

3. The remaining infinitesimals should all be positive except (perhaps) some $*$'s. (The remaining positive infinitesimals may be $+_x$, $0^n \mid +_x$, $\uparrow$, $\Uparrow*$, ...) If there are an odd number of $*$'s, play one of those. Otherwise play on any positive infinitesimal.

4. All that remains are numbers which are explained in Section 2.1.

2.6 Multiple invasions

Although mathematics is helpful in all of the positions discussed so far, it is quite plausible that the strong Go player could play the positions correctly (or at least be taught to do so without reference to any mathematics). However, there are some positions whose values are more subtle. The first *teaser* problem posed in Section 1.3 includes many such positions. The recurring theme is single unconnected groups making multiple invasions as in Figure 2.13. We'll quickly sketch how to evaluate these multiple invasions, but the reader should not feel obliged to understand the method yet.

In order to analyze such a position, first analyze the position as if the invading group were connected as in Figure 2.14. Each blocked corridor is labeled with a b_i, and each unblocked corridor is labeled with a u_j except the shortest unblocked corridor labeled s.

Now, to find the value of the original position (in Figure 2.13), there are three cases:

---

[7]In Appendix E this table is extended to include some kos.

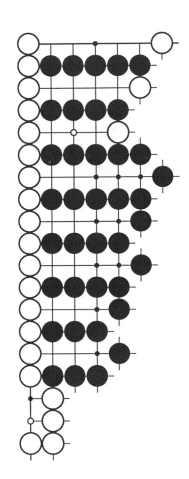

Figure 2.13: *Unconnected group invading many corridors*

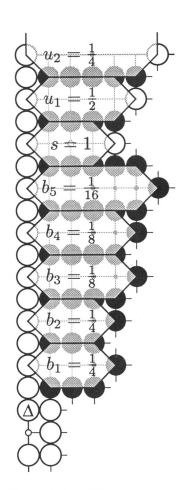

Figure 2.14: *The same group connected at* Ⓐ

1. If the sum of the values of all the blocked corridors is at least 1, then just assume the stone at Ⓐ has in fact been played. Each corridor's value is unchanged.

2. If b_i's add up to something less than 1 and there is an unblocked corridor, s, then an interesting thing happens. The shortest unblocked corridor becomes value $s/2$. Each blocked corridor becomes worth $b_i s/2$, and the remaining unblocked corridors retain their old value.

3. The last possibility is that $\sum b_i < 1$ and there are no unblocked corridors. Then, the position has value $-_x \,|\, 0^n$, where n is one less than the number of blocked corridors, and x depends on the value of capturing the invading group.

The example in Figure 2.13 falls into the second case. Here,

$$\sum b_i = \frac{1}{4} + \frac{1}{4} + \frac{1}{8} + \frac{1}{8} + \frac{1}{16} = \frac{13}{16} < 1$$

So the blocked corridors b_1, b_2, b_3, b_4, and b_5 have values $b_i s/2$ or $\frac{1}{8}$, $\frac{1}{8}$, $\frac{1}{16}$, $\frac{1}{16}$, $\frac{1}{32}$, respectively. Unblocked corridor s has value $s/2$ or $\frac{1}{2}$, and unblocked corridors u_1 and u_2 retain their original values of $\frac{1}{2}$ and $\frac{1}{4}$. Further, these fractions indicate the value of moving: the move on corridor b_5 has value $-\frac{1}{32}$ when playing the chilled game.

The last case is when $\sum b_i < 1$ and there are no unblocked corridors as in the following position:

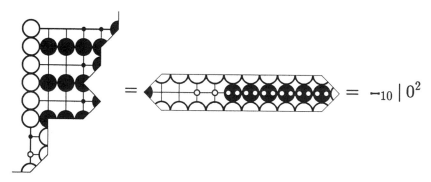

Each Black move approaching the invading group on the right corresponds to a Black move blocking the invasion of a corridor on the left. And the White move blocking the corridor on the right corresponds to White invading each corridor on the left in succession while Black blocks each one until finally White connects:

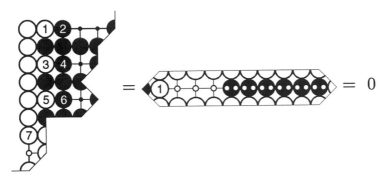

The industrious (or, perhaps, impatient) reader who truly wishes to understand multiple invasions may wish to turn to Sections 4.8 and 4.9, where two related problems (including the teaser) are solved. However, the real purpose in presenting the result so early in the book is to motivate the mathematics and give a flavor of results to come.

Chapter 3

Mathematics of Games

3.1   Common concerns

We will soon present the basics of combinatorial game theory much more formally. But first we'd like to address the most common questions which Go players ask when first introduced to the theory.

Why use mathematical notation? Why do we need $*$, $\uparrow$, $+$, etc.? Aren't the Go positions they represent enough?

Games like warmed $\uparrow$ don't merely represent one position, but rather a rich class of positions. (See, for example, Figure E.4 on page 191.) And the mathematics helps to combine $\uparrow$ with other positions. For example, consider the equation:

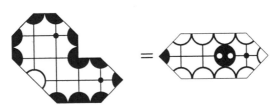

This fact is very hard to come upon by chance. However, suppose you come upon the position on the left, and decide to analyze it using combinatorial game theory. Then, after writing out the possible move sequences and using the methods in this chapter, you can find the chilled position equals $-_1$. Knowing the value is $-_1$ tells you *exactly* how much the position is worth.

Or consider the equation,

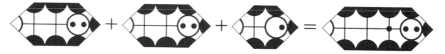

I think you'll agree this is much more difficult to remember and use than

$$\uparrow + \uparrow + * = \Uparrow *,$$

particularly if the values come up in disguised ways. (As usual, you can verify this equation yourself by testing that the second player moves last on the game . In other words, no matter who moves first, Black wins by two points on the unchilled game

.)

The ability to estimate the score is invaluable to the strong Go player. This ability provides a solid foundation for all decisions made in a Go game. A richer vocabulary for describing values of positions and incentives of moves fortifies this foundation. The language of Go literature is simply not precise and rich enough to explain the ideas in this book.

Why is chilling useful?

The first reason chilling is useful is, in a sense, historical. Combinatorial game theory has been used most successfully to analyze *tepid* and *cold* games. Loosely speaking, these are games where the score is the same no matter who moves first. (The object of these games is to move last.) Since one-point Go endgames, when chilled, become tepid games, and since the chilling process can be reversed, we can take advantage of all the successes of combinatorial game theory by using chilling.

Still, one might argue that we could rewrite the entire book without chilling, simply using the same notation (*, ↑, +, etc.) on the unchilled games. For instance, one might define $* = \{1|-1\}$ rather than $\{0|0\}$. Our rules for arithmetic on these games still work just fine. However, this view has a completely different connotation: The emphasis would be on accumulating points rather than gaining the last move.

Another clear advantage with chilling and warming is that the technique is very general. One can cool by more than just one point. The effect, however, is no longer invertible — information about one point differences is lost. This is an important direction of ongoing research, which is mentioned again in Chapter 5. We expect more results in this area to be published in the next few years.

How much of the rest of this chapter is required knowledge?

That all depends on how much you wish to get out of the book. There are several options open to you:

- Understand the chapter thoroughly. Read the chapter with care, and go a little beyond what is written. Take out a pen and paper, and work out the details of some assertions for yourself. You can also gain intuition by playing with the computer program [Wol] mentioned in Section 1.4. You can test your understanding by trying to find the canonical forms of rooms you select from the tables beginning on pages 71 and 202. If anything is still confusing, you may wish to refer to other references on combinatorial game theory — *Winning Ways* [BCG82] or *On Numbers and Games* [Con76].

  After taking these steps, you will have the tools to understand all of the results in the book and their proofs. Most importantly, you'll be able to work out positions for yourself — perhaps specific ones that come up in your games, or from practice problems in Appendix C, or infinite classes of positions you find interesting.

- Believe all the statements made. Read the chapter carefully enough to believe most assertions. The emphasis here is on *belief* rather than *understanding*. You will appreciate most of the statements and proofs in the book (except, perhaps, those which come at the very end of Chapter 4), but may not understand the proofs, and will not be able to apply the ideas to analyze new positions which are not explained in the book. You will, however, be able to do many of the problems in Appendix C.

- Just skim the rest of this Chapter; perhaps you are put off by the mathematical notation. In this case you should read with the goal of reinforcing the key ideas presented in Chapter 2. These ideas are presented again more formally in this and the following chapter. But you will probably want to skip the proofs and also skip Sections 4.6, 4.7, 4.8, 4.10, and 4.11. If you've had enough of combinatorial game theory, you may wish to skip to Chapter 5 on future research possibilities and Appendices A and B about rules.

3.2   Sums of games

Consider the four endgame problems in Figure 3.1: Each problem appears simple enough — all moves are worth around a point and there aren't many possible moves. It's quite reasonable for a strong amateur to find the winning lines by using a few rules of thumb and lookahead. (The techniques introduced in Chapter 2 solve the problems without lookahead.)

The problem in Figure 3.2 appears far more complicated. But the problem has been carefully constructed to simply combine the four smaller endgame problems; Figure 3.1(a) is in the same orientation, (b) and (d) are rotated counterclockwise and (c) is reflected. We'll say that the large problem is the *sum* of the four smaller problems. Each of the four smaller problems is called a *summand*. Using mathematics of games, if you've done the analysis to solve the four smaller problems then you've done enough to solve the large problem. That is, solving the large problem is no harder than solving the four smaller.

The goal of mathematical game theory is to determine just how much information must be kept about a summand to know the outcome of the whole board when played perfectly. The information kept about a particular summand is independent of the rest of the board. For example, using the techniques of Chapter 2, we can evaluate the four positions in Figure 3.2. The values of all regions in the four smaller diagrams are shown in Figure 3.3.

Surprisingly, that's all the information that is needed about each piece to find a winning move for the large problem. In particular, the large problem has value $* + \uparrow* + \downarrow + \uparrow* = \uparrow*$, and a winning move is found on $\uparrow$ moving it to $*$. As a result, without knowing anything about the details of the four positions, there must be winning move(s) in positions (b) and (d), and each of these moves is at least as good for White as converting an $\uparrow$ to a $*$. In this case, the moves are to attack the $\Uparrow$ in (b) or the $\uparrow$ or $\Uparrow*$ in (d). Other winning moves may also exist, but being able to analyze each of the four pieces will indicate at least one of the winning moves.

In summary, the mathematics gives exactly enough information on each summand to be able to (a) determine the outcome of the whole game under optimal play and (b) find a summand with a winning move if one exists.

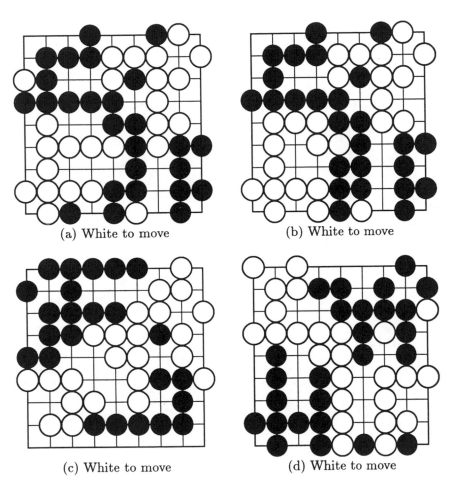

(a) White to move   (b) White to move

(c) White to move   (d) White to move

Figure 3.1: *Four small endgame problems*

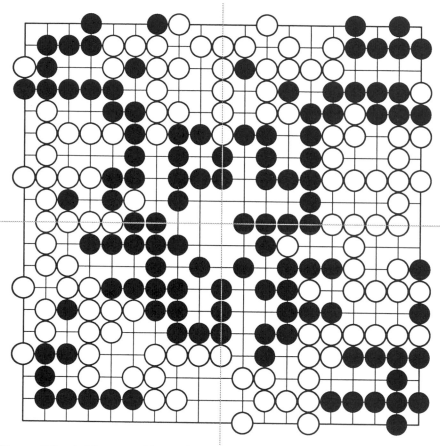

Figure 3.2: *A bigger problem which actually consists of the 4 smaller problems pasted together*

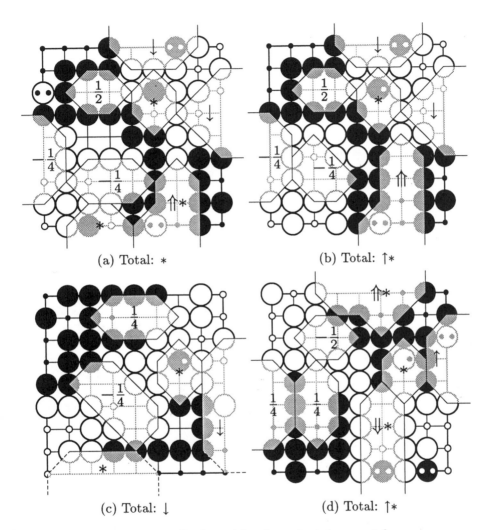

(a) Total: ∗   (b) Total: ↑∗

(c) Total: ↓   (d) Total: ↑∗

Figure 3.3: *Each position is evaluated separately*

3.3   Difference games

Before diving into combinatorial game theory, this section gives a simple motivating example. In Figure 3.4, assume there are no kos on the rest of the board, and further that all stones shown are alive with the exception of the two stones labeled ▲.

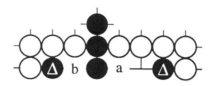

Figure 3.4: *Endgame question*

The question is which move is better for Belle Black — *a* or *b*? Or does it depend on what is on the rest of the board?

First, study the diagram in Figure 3.5(a). The bottom half of the board has the same position as Figure 3.4, and the top has the position with colors reversed. Assuming the rest of the board has been played out and is a tie, the game should be a tie, no matter whose turn it is. The player moving second can just mirror what the other player does. But suppose Black plays at the move corresponding to *a* in Figure 3.4, and White plays at *b*. Then we arrive at the second diagram. We'll call this position $G_a - G_b$, where G_a is the position after Black moves at *b*, and $-G_b$ is the position after Black moves at *b* *except* the negative indicates that the colors are reversed

From this position, White can only attain a tie moving first, but Black can win moving first, as shown in the diagrams in Figure 3.6. Remember, that either player should be able to get a tie moving second in Figure 3.5(a) by maintaining symmetry, and so White's play at ⓑ is a poor response to Black's ⓐ. Hence, there exists at least one position when the move at *a* is better than the move at *b* in Figure 3.4.

In fact, this demonstration is enough to show more. No matter what else is on the board (as long as there are no kos), move *a* is always at least as good as move *b*!

If we treat a tie score as a win for the last player, there are four possible

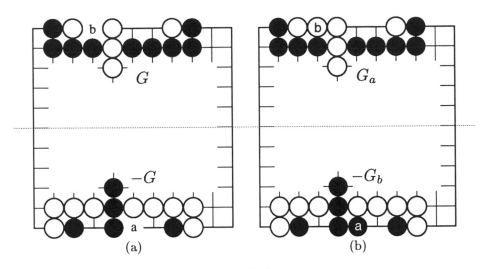

Figure 3.5: *Differencing*

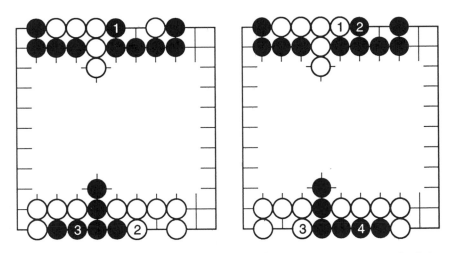

Figure 3.6: *Black wins moving first (left), but White only ties (right)*

outcomes when playing the difference game $G_a - G_b$:

Black wins no matter who moves first	G_a is better than G_b
White wins no matter who moves first	G_b is better than G_a
Second player to move wins	G_a and G_b are worth the same
First player to move wins	Which is better depends on context

3.4   Simplifying games

There are two basic techniques for simplifying games called *deleting dominated options* and *reversing reversible options*. In fact, as we'll see in Section 3.5.2 about canonical forms, these are provably the only two techniques ever needed! After repeated application of the two techniques, any game can be simplified to its unique smallest equivalent game.

3.4.1   Deleting dominated options

The first technique will seem obvious. If a player has two moves available, and one is always better than the other, then there's no point in considering the worse option. For example, suppose we're trying to simplify $\uparrow + *$. Black and White each have two moves available: Black can convert the $\uparrow$ to 0 (leaving $*$), or she can convert the $*$ to 0 (leaving $\uparrow$). White can move $\uparrow$ to $*$ (leaving $* + * = 0$) or he can move $*$ to 0 (leaving $\uparrow$):

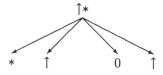

Since White, favoring smaller valued games, prefers 0 to $\uparrow$, we can delete the option to $\uparrow$ without changing the game:

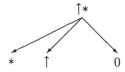

We say that Right's option to ↑ is *dominated by* 0, and so we can delete the dominated option. In this case, we already *knew* 0 < ↑, and so (from White's perspective) ↑ is dominated. Sometimes this can be more subtle. For example,

$b < c < a$

White's move at b is best.

$a < c$, but

a and b are incomparable

White should move at a or b depending on context.

When in doubt, play the difference game!

3.4.2 Reversing reversible options

If Black can make a move for which White has a reply which leaves a position which White prefers to the original, then Black's move is said to *reverse through* White's reply. If a Black's move reverses, it still might be reasonbable for Black to make the move. However, there's little point in doing so unless she plans on continuing to move in the area immediately after White replies. (Throughout most of the book we're assuming there are no ko's on the board, so Black's first move is not a ko threat.)

In other words, if a Black move to A reverses through a White move to A^R, then the game can be simplified by replacing Black's move(s) from A^R for Black's move from G:

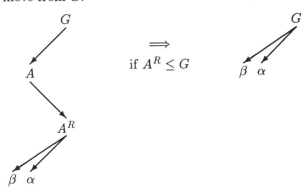

If there are other moves available from G, they remain. Other available moves from A, whether they be Black's or White's, vanish entirely! White's moves from A^R also vanish. In summary, the thin lines all disappear:

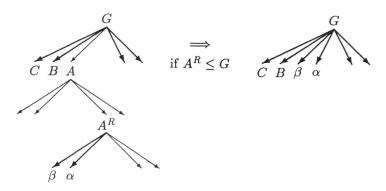

As an example, let's return to the example which we began to simplify in the last section:

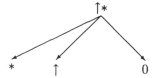

If Black moves on the $*$ in $\uparrow + *$, that would leave the position $\uparrow$. If White were to then move this new position to $*$, then White would be better off than he started: $* < \uparrow *$. So Black's move to $\uparrow$ reverses through White's move to $*$, and the position simplifies:

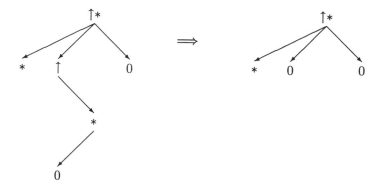

This last game, $\{0, *|0\}$, is simplified as much as possible, and is called the *canonical form* $\uparrow*$. The canonical form for a game is unique even though there may be many different alternatives for simplifications which arrive at this canonical form.

Reversible moves may seem a bit mysterious, but Go players instinctively reverse out moves all the time. For example, in the position

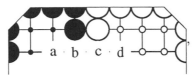

White's move at b reverses through Black's move at a to c. Similarly, Black's move at c reverses through White's move at d to b. Hence,

$$= \{0|0\} = *$$

In short, as most Go players would agree, the move is worth two points in gote. However, another simple example should demonstrate that it's possible to draw some counterintuitive conclusions using reversibility. In the position

it's not hard to see that Black's best move is at b. However, White's plays at a and b are incomparable. The game tree for the position is shown in Figure 3.7. In short, the position's worth is given by

$$= -1 \mid *, \downarrow$$

It turns out that *both* of White's moves reverse. The move to reverses through , but from the latter position White has *no* move.[1] So

---

[1] For any singly marked captured stone, , in Figure 3.7, the marking cancels the value of the capture, so we omit such marked stones in this paragraph.

You may want to use differencing to check for yourself that $- \geq 0$.

Or you can symbolically verify that $0 - \{-1 \mid *, \downarrow\} \geq 0$, i.e., $\{-1 \mid *, \downarrow\} \leq 0$.

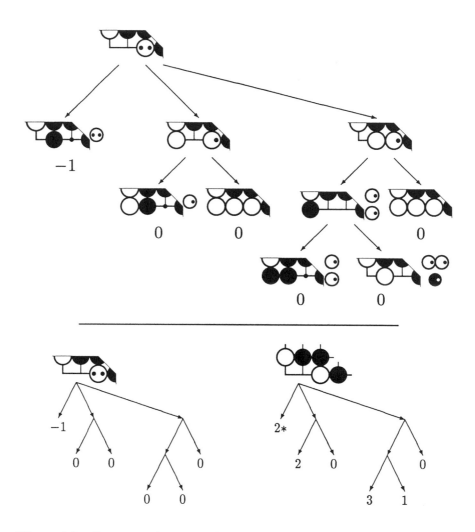

Figure 3.7: *Game tree for example position. In the lower left, the same tree is repeated, and in the lower right the game is shown unchilled*

the option to move to disappears entirely! In other words, after the exchange, Black's position has improved, so White should move to only if he plans to continue moving from , which would always be ill-advised. Due to the reversibility,

$$-1 \mid *, \downarrow = -1 \mid \downarrow$$

Also, reverses through to , and therefore $-1 \| \downarrow = -1 \mid 0 = -\frac{1}{2}$. Hence,

$$= \{-1 \mid 0\} = -\frac{1}{2}$$

3.5 Combinatorial game theory

This section briefly summarizes the basic definitions of mathematical games and some of the key results about them. Many of the ideas have already been introduced at least once, but this presentation is more formal, expanding and generalizing the key points. Although extremely terse and dense, the section provides enough facts to understand the remainder of the book. However, you may wish to reread the section several times as you read the book in order to understand the implications. And don't hesitate to take out a pencil and paper and test the claims for yourself! Or try playing with the computer program [Wol] discussed in Section 1.4. For more detail, refer to *Winning Ways* [BCG82] or *On Numbers and Games* [Con76]. Also, an introduction to the applications to Go can be found in [Ber91].

3.5.1 Definitions

A game, G, between two players, Left and Right, is defined as a pair of sets of games,[2]

$$G = \left\{ \mathcal{G}^L \mid \mathcal{G}^R \right\}$$

---

[2]In Section 3.5.1 only, we have written G for a single game and $\mathcal{G}$ for a set of games. In accordance with the literature, we will quickly abandon this distinction, writing simply G for both a set and a single game.

The interpretation is that if it is Left's move, she can move to any game $G^L \in \mathcal{G}^L$, and it becomes Right's turn. If a player, on her turn, cannot move (e.g., $\mathcal{G}^L$ is the empty set on Left's move), that player has lost. The *followers* of G are the elements of $\mathcal{G}^L$ and $\mathcal{G}^R$. The *positions* of a game G are G itself and all positions of followers of G. A few things are worth noting:

- The reader may be worried that the lack of a concept of score will preclude Go from being a game. Score will fall out of the theory naturally, so don't panic!

- A game has a dual meaning — its common usage as a set of rules, and its technical definition used here, which represents the game tree of a specific position.

- Whose turn it is to move is not part of a game's definition.

- In this discussion all games are finite and loopfree. (Games involving kos entail additional complications which are not covered in this volume, except for a superficial treatment of one-point kos in Section 5.3.)

We will abbreviate games by removing the {}'s whenever possible:

$$G = \left\{ \{G_1^L, G_2^L, \ldots\} \mid \{G_1^R, G_2^R, \ldots\} \right\}$$
$$= \{G_1^L, G_2^L, \ldots \mid G_1^R, G_2^R, \ldots\}$$

We write more |'s to indicate a higher level in the game tree. For example,

$$\left\{ G_1 \mid \{G_2 \mid G_3\} \right\} = G_1 \| G_2 \mid G_3$$

The negative of a game is defined by reversing the roles of the players:

$$-G \overset{\text{def}}{=} \left\{ -\mathcal{G}^R \mid -\mathcal{G}^L \right\},$$

where $-\mathcal{G}^R \overset{\text{def}}{=} \{-G^R\}_{G^R \in \mathcal{G}^R}$. A sum of games is given by

$$G + H \overset{\text{def}}{=} \left\{ (\mathcal{G}^L + H) \cup (G + \mathcal{H}^L) \mid (\mathcal{G}^R + H) \cup (G + \mathcal{H}^R) \right\},$$

where $\mathcal{G}^L + H \overset{\text{def}}{=} \{G^L + H\}_{G^L \in \mathcal{G}^L}$. The interpretation is that G and H are played side by side, and a player on his move can move on either G or H. Once again, if a player cannot move on either G or H, (e.g., on Right's move, $\mathcal{G}^R$ and $\mathcal{H}^R$ are both empty), the player loses. When unambiguous, we will often write GH instead of $G + H$, just as grade school children are taught to write $1\frac{1}{2}$ to mean $1 + \frac{1}{2}$.

We'll say $G_1 \geq G_2$ if, for all H, Left does at least as well on $G_1 + H$ as on $G_2 + H$. If Left wins on $G_2 + H$ moving first (second), then she wins moving first (second) on $G_1 + H$. We'll say $G = H$ if $G \geq H$ and $H \geq G$.

Under these definitions, equivalence classes of finite games form a mathematical group under $+$, with a partial order $\geq$, where a zero game is a game where the second player on move can win (for example, $G = \{|\}$). Furthermore, $G \geq 0$ if and only if Left can win moving second on G. So to test if $G_1 \geq G_2$, we can test if $G_1 - G_2 \geq 0$, i.e., that if Left can win moving second on $G_1 + (-G_2)$.

So comparing G and H yields four outcome classes, the last being when G and H are incomparable:

$$
\begin{array}{lll}
G > H & G - H > 0 & G - H \text{ is a win for Left} \\
G = H & G - H = 0 & G - H \text{ is a win for second player} \\
G < H & G - H < 0 & G - H \text{ is a win for Right} \\
G \parallel H & G - H \parallel 0 & G - H \text{ is a win for first player}
\end{array}
$$

(The symbols $\parallel$ and $\not\gtrless$ are both used in the literature to denote incomparable games.)

3.5.2 Canonical forms

Conway discovered that for any game G, there exists a unique minimal (i.e., smallest game tree) game G' such that $G = G'$. G' is called G's *canonical form* and can be obtained by repeated application of the following two operations on G and its positions. Let

$$G = \{A, B, C, \ldots \mid D, E, F, \ldots\}$$

Deleting dominated options: If $B \geq A$, then B is said to *dominate* A, and

$$G = \{B, C, \ldots \mid D, E, F, \ldots\}$$

Similarly, if $E \leq D$, then E dominates D and

$$G = \{A, B, C, \ldots \mid E, F, \ldots\}$$

Reversing reversible options: If A has a right option, A^R, with $A^R \leq G$, then Left's move to A is said to *reverse* through A^R, and A can be replaced by Left's options from A^R in G. If $A^R = \{\alpha, \beta, \ldots \mid \gamma, \delta, \ldots\}$, then

$$G = \{\alpha, \beta, \ldots, B, C, \ldots \mid D, E, F, \ldots\}$$

Similarly, if there exists $D^L \geq G$, Right's move to D reverses.

If none of the positions of a game have any dominant options nor any reversible options, then the game is said to be in *canonical form*. If two games in canonical form are equal in value, they must also be identical in form [BCG82, p. 78].

On a sum of games, if Left can win, Left can still win by pretending all of the summands are in canonical form, and she'll still find a winning move. Some other winning moves might be obscured, however, by the conversion to canonical form.

3.5.3   Examples of games

The concept of a score falls naturally out of the theory. For positive integers n and positive dyadic rational $\frac{m}{2^k}$ (m odd) we define:

$$0 \;\overset{\text{def}}{=}\; \{\mid\}$$

$$n \;\overset{\text{def}}{=}\; \{n-1 \mid \}, \quad n > 0$$

$$\frac{m}{2^k} \;\overset{\text{def}}{=}\; \left\{\frac{m-1}{2^k} \,\middle|\, \frac{m+1}{2^k}\right\}$$

For example,

$$\frac{1}{2} \;=\; \left\{0 \,\middle|\, 1\right\}$$

$$\frac{1}{4} \;=\; \left\{0 \,\middle|\, \frac{1}{2}\right\}$$

$$\frac{3}{4} \;=\; \left\{\frac{1}{2} \,\middle|\, 1\right\}$$

Arithmetic on these games works analogously to standard arithmetic on numbers. The *Number Avoidance Theorem* [BCG82, p. 145] states that if G is a number and H is not, then optimal play on $G + H$ is given by playing on H. Therefore, one can compress the canonical form into a shorter tree which has numbers at its leaves. Each number can be thought of as a score for the game, a positive score being better for Left. (One might be tempted to define $\frac{1}{2} = \{1|0\}$ since it also satisfies $\{1|0\} + \{1|0\} = 1$, but this fails to give the Number Avoidance Theorem which plays an important role in the Theory.)

Another important fact is that all games of the form $g = \{a|b\}$, for $a < b$ numbers, are also numbers. In particular, g is the *simplest number* between a and b. If there are integer(s) between a and b, g is the integer of smallest absolute value. Otherwise, g is the dyadic rational, $i/2^j$, of smallest denominator between a and b. So, for example,[3]

$$\left\{\frac{5}{8} \;\middle|\; \frac{15}{16}\right\} = \frac{3}{4}$$

The games in Figure 3.8 are *infinitesimals*, exceeding all negative numbers and less than all positive numbers. Although they don't come up directly as Go positions, they are well understood in the theory, and can be mapped by chilling to many one point endgame positions. When they differ, a game's definition is followed by its canonical form.

The following facts can be verified. You are encouraged to work out these facts by hand, as well as to reduce some of the sums above (such as $\uparrow + *$ and $\uparrow + \uparrow$) to canonical form.

- $*$ is incomparable with 0, $+_r$, or $\uparrow$.

- For $n \geq 2$, $n \cdot \uparrow \; > *$

- For $r_1 > r_2 > 0$ rationals, $\uparrow > +_{r_2} > +_{r_1} > 0 > -_{r_1} > -_{r_2} > \downarrow$

A game like Go is *hot*. Unlike numbers and infinitesimals, players typically gain more points by moving. The *left stop* is defined to be the maximum number Left can reach by moving first if Right plays perfectly. More technically, left stop and right stop are defined recursively in tandem:

$$\text{LS}(G) = \begin{cases} G & \text{if } G \text{ is a number} \\ \max\{\text{RS}(H) : H \text{ a left option of } G\} & \text{otherwise} \end{cases}$$

$$\text{RS}(G) = \begin{cases} G & \text{if } G \text{ is a number} \\ \min\{\text{LS}(H) : H \text{ a right option of } G\} & \text{otherwise} \end{cases}$$

---

[3] Conway also discovered a more general class of numbers, called surreal numbers [Con76] [Knu74]. Surreal numbers solve infinite games, and provide a modern framework to unify the work of classical set theorists such as Dedekind and Cantor. But all numbers used in this book have finite birthdays, and values which are dyadic rationals.

$*$	$\overset{\text{def}}{=}$	$\{0\|0\}$	star
$\uparrow$	$\overset{\text{def}}{=}$	$\{0\|*\}$	up
$\downarrow$	$\overset{\text{def}}{=}$	$-\uparrow$	down
$\uparrow*$	$\overset{\text{def}}{=}$	$\uparrow + *$	up-star
	$=$	$\{0,*\|0\}$	
$n\cdot\uparrow$	$\overset{\text{def}}{=}$	$\overbrace{\uparrow+\uparrow+\ldots\uparrow}^{n},\ n\geq 2$	double-up,
	$=$	$\{0\ \|\ (n-1)\cdot\uparrow + *\}$	triple-up, ...
	$=$	$\{0\ \|\ (n-1)\cdot\uparrow*\}$	
$n\cdot\uparrow*$	$\overset{\text{def}}{=}$	$n\cdot\uparrow + *,\ n\geq 2$	double-up-star, ...
	$=$	$\{0\ \|\ (n-1)\cdot\uparrow\},\ n\geq 2$	
$\Uparrow$	$\overset{\text{def}}{=}$	$2\cdot\uparrow$	
$+_G$	$\overset{\text{def}}{=}$	$\{0\|\|0\|-G\},\ G > r,\ r > 0$ a number	tiny G
$-_G$	$\overset{\text{def}}{=}$	$\{G\|0\|\|0\},\ G > r,\ r > 0$ a number	miny G

Figure 3.8: *Definitions of common infinitesimals*

An infinitesimal has left stop and right stop both zero. A *hot* game has left stop exceeding right stop. A *cold* game is a number. A game which differs from a number by a non-zero infinitesimal is said to be *tepid*. Any game is either cold, tepid or hot.

3.5.4 Cooling

$G = \{G^L|G^R\}$ *cooled* by the non-negative number t [BCG82, pp. 147–149], denoted G_t, is defined by:[4]

$$G_t = \left\{G_t^L - t \ \middle|\ G_t^R + t\right\},$$

unless for some $\tau < t$, $\{G_\tau^L - \tau|G_\tau^R + \tau\}$ is infinitesimally close to a number x, in which case

$$G_t = x$$

Cooling by t is a "tax" placed on the mover. The mover pays t points for the privilege of moving. The tax can be sufficiently high that it will be

---

[4] As promised by the footnote on page 45 we are now using G to denote both a single game and a set of games.

disadvantageous to move — when $t > \tau$. G is said to have *temperature* τ, *mean* x, and *freeze* to G_τ. Conway established the following important properties:

- Linearity: $G_t + H_t = (G + H)_t$

- Order preserving: $G \geq H \Rightarrow G_t \geq H_t$

- $\text{mean}(G + H) = \text{mean}(G) + \text{mean}(H)$

- $\text{temperature}(G + H) \leq \max(\text{temperature}(G), \text{temperature}(H))$

3.5.5 Incentives

The notion of "how much a move is worth" is encapsulated in *incentives* [BCG82, p. 144]. The elements of the set

$$\Delta^L \{G\} \quad \overset{\text{def}}{=} \quad G^L - G$$
$$\overset{\text{def}}{=} \quad \{H - G : H \in G^L\}$$

are the left incentives of G. These indicate the value of a move to Left. The right incentives are given by

$$\Delta^R \{G\} \quad = \quad G - G^R$$

The sign of the last definition was chosen so that both Left and Right are eager to make moves with large incentives. Lastly, the incentives of a game are defined by

$$\Delta \{G\} = \Delta^L \{G\} \cup \Delta^R \{G\}$$

Incentives are dependent on the form of the game. For instance,

$$\Delta^L \left\{ \tfrac{1}{4} \middle| 1 \right\} = -\frac{1}{4} \neq -\frac{1}{2} = \Delta^L \{0|1\}$$

even though $\tfrac{1}{4}|1 = \tfrac{1}{2} = 0|1$. Sometimes we call these *formal incentives*, and define *canonical incentives* as incentives computed from the canonical form. Thus, the game $\tfrac{1}{4} \middle| 1$ has formal left incentive $-\tfrac{1}{4}$ and canonical left incentive $-\tfrac{1}{2}$.

3.6 Warming

Cooling a game of Go by one point nearly has an inverse, which is given by
a *warming* operator. A warming operator is any function which *overheats*
from s to t, where s and t are both infinitesimally close to 1. Overheating,
denoted $\int_s^t$, was introduced by Berlekamp [Ber88] to analyze Blockbusting
games and has also been used to simplify Domineering positions in [Wol93].
Although the reader need not be concerned with its general definition, it is
provided in the glossary.

In Go, the warming operator which inverts cooling by one point is $\int_{1*}^1$,
and is equivalent to a *Norton Multiply* [BCG82, p. 246] by 1∗. As a consequence of being a Norton Multiply, this warming operator is linear and
order preserving. Since $\int_{1*}^1$ is the only warming operator used in this book,
we will abbreviate it simply $\int$. In short, for a game $G = \{G^L | G^R\}$,

$$\int G = \begin{cases} G & \text{if } G \text{ is an even integer} \\ G* & \text{if } G \text{ is an odd integer} \\ \{1+\int G^L \mid -1+\int G^R\} & \text{otherwise} \end{cases}$$

This last equation should be understood in both directions. To compute
heated values one uses it in the forward direction, but often when working
out problems in a bottom up fashion one uses the equation in the opposite
direction.

A Go position is *even* (or *odd*) if the number of empty intersections plus
the number of prisoners captured is even (or odd). (Notice that removing
an effectively captured stone from the board and adding it to the prisoners
has no affect on the parity.) The parity of a position has the following two
properties:

1. Parities add as expected — the sum of two even (or two odd) positions
   is even, and the sum of an odd and an even position is odd.

2. Parity alternates during play — all followers of an even position are
   odd and all followers of an odd position are even.

We call a Go position *elementary* if, when completely played out in any
environment, every point on the board is either occupied by a live stone, or
becomes territory for a player.[5] The experienced Go player will realize that

---

[5] By *played out*, we mean that all dame are filled, and dead stones are removed. Defining
precisely what is dead and alive at the end of the game can be troublesome, but since it
is a distraction to the main results, it will not be discussed here. It will, however, be clear
in the specific positions discussed.

a game is elementary if none of its canonical positions has any kos or sekis.

For the remainder of Section 3.6, we assume that our game G satisfies the following restriction:

> Let G be an even elementary Go position in canonical form.

We've just argued the following Lemma holds:

Lemma 1.1 *A stopping position of G is even if and only if its value is even.*

Theorem 1 $G = \int G_1$.

Proof: We'll introduce the transformation $f(G) = \int^{-1} G$ which *chills* G,[6]

$$
f(G) \overset{\text{def}}{=}
\begin{cases}
n & \text{if } G \text{ is of the form } n \text{ or } n* \\
\left\{ f\left(G^L\right) - 1 \;\middle|\; f\left(G^R\right) + 1 \right\} & \text{otherwise}
\end{cases}
$$

This is nearly the same as cooling by one point, except we ignore when the exceptional case occurs, when G_τ is a number for $\tau < 1$. The following series of Lemmas 1.2–1.6 show that (1) $G = \int f(G)$ and (2) $f(G) = G_1$. These two facts combined prove the Theorem. ∎

Lemma 1.2 *G and each of its followers either has left stop exceeding right stop, or is of the form n or $n*$ (n integer).*[7]

Proof: Suppose to the contrary, and that G's left and right stops are both equal to n.

If n is even, then $G = n$. Since G is even, a move to n on G must occur after an even number of moves, and so the second player wins $G - n$. (Moves on n need not be considered by the Number Avoidance Theorem.)

---

[6] In mathematical game theory outside of this book, this is formally identical to *unheating*

[7] $G + H$ is often written simply GH. Number parts are listed first, so $2* = 2 + *$ and $2{\uparrow}3 = 2 + {\Uparrow}$.

If n is odd, then $G = n*$. Consider playing the difference game $G - n*$. Two moves in a row by Left on G can only lead to a stop greater than or equal to n. Hence, the second player can guarantee a win by always playing on G until its stop. If the stop is equal to n, an odd number of plays were made since G is even.

A parallel argument made about subpositions of G which are odd completes the proof.   ∎

Lemma 1.3 $G = \int f(G)$.

Proof: If G is of the form n or $n*$, the lemma holds trivially. Otherwise, it is sufficient to show that $f(G)$ is in canonical form, and hence warming inverts the transformation.

First we'll show $f(G)$ has no dominated options. Suppose, to the contrary, G has two left options, A and B, with $f(A) \geq f(B)$. By Lemma 1.2, left stops exceed right stops. Hence, optimal play on the game $A - B$ is never to play on a leaf of value n or $n*$ until both games are settled. We'll show Left can win on $A - B$ by pretending she's playing $f(A) - f(B)$. Left wins moving second on $f(A) - f(B)$, and we consider two cases:

1. If an even number of moves were made to get to a stopping position, the single point adjustments in the definition of f cancel. Further, since G is even, if the value is exactly 0, the corresponding stopping position in $A - B$ is also 0. Hence $A - B \geq 0$.

2. Similarly, if an odd number of moves were made to get to a stopping position, the stop must be at least 1 since Left won. In the extreme it's 1, and since G is even, the corresponding stopping position on $A - B$ is $*$, and Left can make the last move to zero.

In either case, $A - B \geq 0$, and G was not in canonical form. The similar argument shows that $f(G)$ has no reversible options, and hence $f(G)$ is in canonical form.   ∎

Lemma 1.4 *Let $t = 1/2^i$ and H be a game all of whose stopping positions are multiples of $2t$. Then either H has temperature zero, or H has temperature at least t. Further, H_t's stopping positions are multiples of t.*

Proof: The lemma follows from the observation that for any subtree, I, of H, the distance from I's mean value to a left or right stop is at most its temperature. ∎

Lemma 1.5 *If G has mean $i/2^j$ for i odd, then G's temperature is at least $1 - 1/2^j$.*

Proof: Follows from repeated application of Lemma 1.4 by cooling G first by $\frac{1}{2}$, then cooling it further by $\frac{1}{4}$, etc., through $\frac{1}{2^j}$. ∎

Lemma 1.6 $f(G) = G_1$.

Proof: Assume G is a minimum sized (depth of game tree) counterexample. G's followers satisfy $f(G) = G_1$, and G's temperature, say t, must be less than 1. Let $G_1 = i/2^j$, for i odd. By Lemma 1.5, $t \geq 1 - 1/2^j$. G_t must have a left option with right stop $i/2^j$. Hence, the corresponding stopping positions in $f(G)$ must either be $i/2^j$ with Right moving last, or be at least $(i-1)/2^j$ with Left moving last. Either way, this gives a winning response for Left on the difference game $f(G) - G_1$ when Right moves $-G_1$ to $-(i-1)/2^j$. But Left's options from $f(G)$ are less than or equal to those from G_1, and so $f(G) = G_1$. ∎

Corollary 1 *In elementary Go positions, cooling by one point is the inverse of warming, up to a dame:*[8]

$$\int G_1 = \begin{cases} G \\ \text{or} \\ G* \end{cases}$$

Proof: If G is even, the Theorem applies directly. If G is odd, then note that for games whose left stop exceeds right stop, $\{G^L|G^R\}* = \{G^L*|G^R*\}$. Hence, $G + *$ is even, and $\int G_1 = \int (G + *)_1 = G*$. ∎

If a Go position's game tree has odd seki's (but still no kos) then $*$'s may also show up in subgames of $\int G_1$.

---

[8] A *dame* is Go terminology for the game $*$.

Chapter 4

Go Positions

4.1 Conventions

Belle Black is always the positive, or Left player. Wright White is the negative, or Right player. On Go diagrams, a line running off the board will connect a stone to life. We call stones connected to such lines *immortal*, because they can never be killed or captured. We will typically put small black and white *markings* to indicate an integral number of points we are assigning to the black and white players. These circles might be placed on stones (which could be captured), or on intersections. (No diagrams will have hoshi points — the black dots indicate black points.[1]) Further, the chilling of a game may be indicated by shading and outlining the region. So,

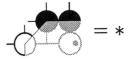 $= *$

Since one white stone may be captured, the game tree is $\{2|0\}$. Now, the black dot indicates black has already been allotted one point for the region, so subtracting that point out leaves $G = \{1|-1\}$. Some arithmetic on games yields that G cooled by 1 is $*$, and in fact, $\{1|-1\} = \int *$.

Warning: The reader may require some time to get used to the markings. A black marking will add a point to White when chilling and marking. It indicates a point that Black has earned in the unchilled and unmarked game, which is set aside for the moment.

---

[1] Go boards typically have small black dots (hoshi points) on several intersections on the board to provide visual cues.

Often we will show a chilled game by simply clipping around the outside stones. In this case, any stone clipped is immortal. For example,

$= \dfrac{1}{2}$

Once again, the value of the game before chilling is $\int \frac{1}{2} = \{1|*\}$

We define

$$0^n|G \;=\; \begin{cases} 0\|0^{n-1}|G & \text{if } n > 0 \\ G & \text{if } n = 0 \end{cases}$$

$$G|0^n \;=\; -\{0^n|-G\}$$

For example, notice

$$0^n|1 \;=\; \dfrac{1}{2^n}$$

$$0^n|0 \;=\; \begin{cases} (n-1)\cdot\uparrow & \text{for } n \text{ even} \\ (n-1)\cdot\uparrow\! * & \text{for } n \text{ odd} \end{cases}$$

$$0^2|-G \;=\; \mathbf{+}_G \quad \text{if } G \text{ exceeds some positive number}$$

4.2 A problem

We have now presented enough machinery to solve the two (closely related) 9×9 endgame problems in Figure 4.1. The diagrams in Figure 4.2 highlight the important regions and include markings on the boards. The dashed lines bound regions which are not obviously independent of their surroundings. For now we'll pretend they're independent, and later verify that the dependencies don't affect the result. In this case, the numbers of black and white markings on each board both equal to 10.

We'll analyze each region using a technique we'll refer to as *playing the chilled game*. Rather than argue about the Go game trees, we'll argue on the chilled regions, and exploit the fact the warming inverts chilling. Since on the chilled game, each move by a player gives up a point, we'll require that each move by, say, White, either removes a black marking or adds a white marking to the board.

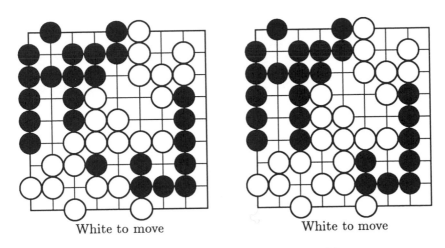

Figure 4.1: *Two related endgame problems*

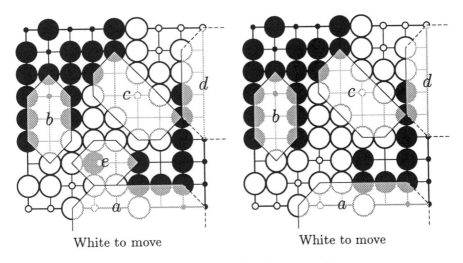

Figure 4.2: *Two related endgame problems*

For example, region e is worth:

$$= \{0|0\} = *$$

(Right's move captures the black stone, although it's been left in the diagram and marked twice — once for the territory and once for the prisoner.)

Region b is also straightforward:

$$= \{0 \| 0 | 1\} = \frac{1}{4}$$

For example, in the position after Right has moved twice, two adjustments have been made to the markings. In the unchilled game $G^{RR} = *$, chilling yields 0, and the markings adjust to 1.

Before looking at region c, notice that Lemma 1.2 on page 53 allows us to conclude that  $= 1$. The unchilled, unmarked game has value $*$, since the game is odd and its left and right stops are both zero. Region c is then given by

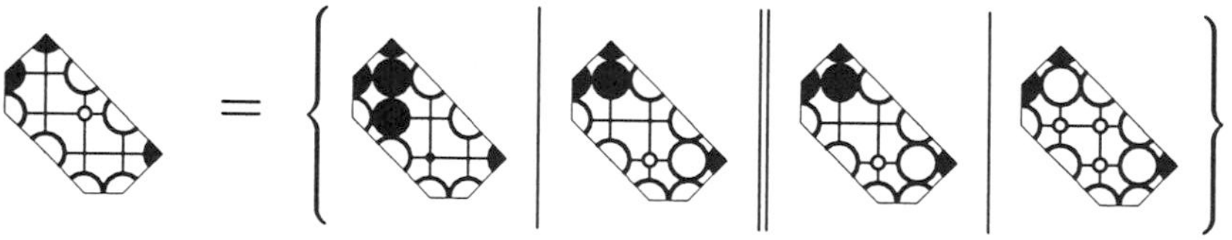

$$= \left\{-1 \middle| -\frac{1}{4} \middle\| -\frac{1}{4} \middle| 0\right\} = \left\{-\frac{1}{2} \middle\| 0\right\} = -\frac{1}{4}$$

Referring back to region a,

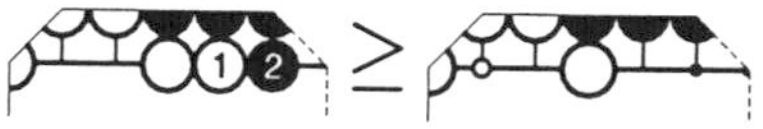

since Left's threat to capture two stones after ① and ❷ are played exceeds any left option before the stones are played. Therefore, Right's move at ①

reverses through left's move at ❷ to

Also, using Lemma 1.2 on page 53,

$$\text{[diagram]} = 0$$

Hence,

$$\text{[diagram]} = \left\{ \text{[diagram]} \,\middle|\, \text{[diagram]} \,\middle\|\, \text{[diagram]} \right\}$$

$$= \{0 \mid 0 \parallel 0\} = \downarrow$$

Lastly, in region d,

$$\text{[diagram]} \geq \text{[diagram]}$$

and moves by either player reverse, so

$$\text{[diagram]} = \left\{ \text{[diagram]} \,\middle|\, \text{[diagram]} \right\} = \{0 \mid 0\} = *$$

So, assuming the regions are independent (an assumption which will be justified shortly), values of the various regions are given in Figure 4.3. The chilled position on the left is now worth $1/4 - 1/4 + * + * + \downarrow = \downarrow < 0$, so Wright can win moving first. Since the markings match, this will give a win for Wright. Wright's move on $*$ would shift the total to $\downarrow *$, which is a win for the first player, and so Belle would get a tie. Wright's move on a number would move the game to $1/4\downarrow > 0$, and again Belle would win the chilled game, getting a tie in the Go game. However, Wright's move on $\downarrow$ would move the game to 0, which is a win for the second player (White), and is Wright White's only winning move.

The chilled position on the right is worth a total of $\downarrow *$ which is a win for the first player. But in this case the winning move for Wright is different. A move on $\downarrow$ to 0 would bring the total to $*$ which is incomparable with 0 and Black can win. However, the move on $*$ would bring the total to $\downarrow < 0$, which White can win.

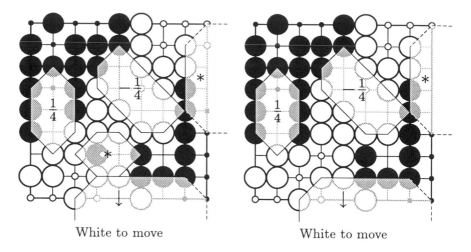

White to move   White to move

Figure 4.3: *Two related endgame problems*

To justify the independence of the various regions, a more careful analysis of each argument will reveal that canonical play is unaffected by the surroundings. In particular, the reversibility in regions a and d still holds, and all stones which were assumed immortal never get captured.

An argument justifying the independence of regions can often be more subtle. The problem of Section 2.4 had an upper right corner which looked something like Figure 4.4.

$$G =$$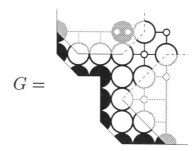

Figure 4.4: *The independence of the two regions needs justification*

It appears as though after multiple moves by Belle, the large white group is susceptible to capture and so the regions are not independent. In particular, five consecutive moves by Black capture White's stones:

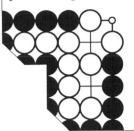

If the white stones were all immortal (uncapturable), the regions of G would each be worth $\downarrow$. Further, the potential for capture of the White group only improves the situation for Black. Hence,

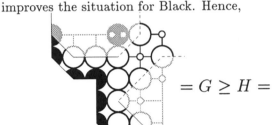 $= G \geq H =$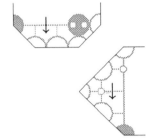

Notice $G^R = H^R$ since both moves by White on G lead to life for all the white stones. Left's move on $H = \downarrow + \downarrow$ to $\downarrow*$ reverses through $*$ to 0. Consequently,

$$H = \Downarrow = \downarrow + \downarrow = \{\downarrow*|0\}$$

Further,

$G^{LR} =$ 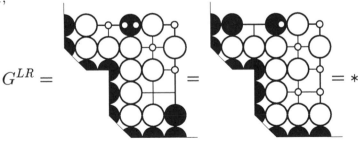 $= *$

since both lines lead to life for the White group. Hence,

$$G \geq H \geq H^{LR} = G^{LR}$$

and either of Left's moves from G to G^L are reversible through G^{LR} as in H. This fact, along with $G^R = H^R$ yields

$$G = H$$

and the independence of the regions in Figure 4.4 is justified.

This *effective* independence can frequently be checked by first assuming independence, and later verifying it. The frequent applicability of this technique demonstrates the remarkable robustness of the results in this book.

4.3   Corridors

Here are some examples of positions which are precisely analyzable, given there are no kos on the board.

Theorem 2 *A chilled "blocked corridor" of length n with $n-2$ markings is worth 2^{1-n} ($n \geq 1$).*

Here are some examples of *blocked corridors*. Recall (from Section 4.1) that the $n-2$ dots or markings on the following board positions normalize their chilled values. Therefore, the unmarked, unchilled values of the positions are $n - 2 + \int 2^{1-n}$.

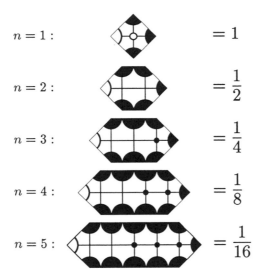

Proof: The proof will be by playing the chilled game. First, fix the markings (dots) for a piece of solid territory by marking all enclosed points, e.g.,

To verify the Theorem when $n = 1$, observe $-1 + \int 1 = *$. When $n \geq 2$, a move by either player adjusts the dots or markings in the diagrams by one in the player's favor. So, observing that $n + \int G = \{n + 1 + \int G^L \mid n - 1 + \int G^R\}$, we need only check that the chilled and marked values are consistent with the Theorem. This example shows $n = 4$:

$$\langle\!\!\langle \cdots \rangle\!\!\rangle = \left\{ \langle\!\!\langle \cdots \rangle\!\!\rangle \,\middle|\, \langle\!\!\langle \cdots \rangle\!\!\rangle \right\}$$

$$= \left\{ 0 \,\middle|\, 2^{1-(n-1)} \right\} = 2^{1-n}$$

∎

The next Theorem concerns corridors of territory invaded at both ends.

Theorem 3 *A chilled "unblocked corridor" of length n with $n - 4$ markings is worth 2^{3-n} $(n \geq 2)$:*

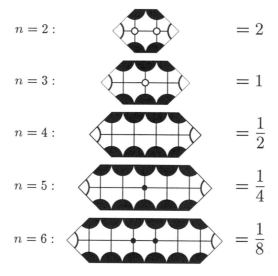

$$n = 2 : \qquad\qquad\qquad\qquad = 2$$

$$n = 3 : \qquad\qquad\qquad\qquad = 1$$

$$n = 4 : \qquad\qquad\qquad\qquad = \frac{1}{2}$$

$$n = 5 : \qquad\qquad\qquad\qquad = \frac{1}{4}$$

$$n = 6 : \qquad\qquad\qquad\qquad = \frac{1}{8}$$

Proof: To check the base case when $n = 2$, observe $-2 + \int 2 = 0 = \{*|*\}$. Once again, when $n \geq 3$, moves by either player adjust the markings in their favor by one, so we can play the chilled game. A move by Left (or Right) on an unblocked corridor of length n moves to a blocked (or unblocked) corridor of length $n - 1$. $\left\{2^{1-(n-1)} \mid 2^{3-(n-1)}\right\} = 2^{3-n}$, since 2^{3-n} is the simplest number between the options. ∎

Theorem 4 *A chilled corridor of length n leading to an x point gote[2] play with $x + n - 1$ markings is worth $0^{n+1} \mid 2-x$.*

For example when $n = 2$ and $x = 4$,

$= \{0 \mid\mid\mid 0 \mid\mid 0 \mid -2\} = 0\mid+2$

Proof: (omitted) ∎

So the unchilled game is worth $x + n - 1 + x + n - 1 + \int 0^{n+1} \mid 2-x$. A special case of this is that a two point gote play is worth $1 + \int *$, and that a corridor of length n leading to a two point gote play is worth $1 + n + \int n \cdot \uparrow$ for n odd, and $1 + n + \int n \cdot \uparrow *$ for n even:

$n = 0$: $= *$

$n = 1$: $= \uparrow$

$n = 2$: $= \Uparrow *$

$n = 3$: $= \Uparrow\uparrow$

---

[2] An x point gote is Go terminology for any game of the form $0\mid-x$ plus an integer.

4.4   Sums of corridors

Adding and subtracting games of the form $x|0^n$ for varying x can lead to complicated canonical forms. However, we can still say a lot about these games. First, Figure 4.5 gives some clues as to optimal play. Although the chart only partially orders the incentives, the following Theorem shows that if games of this form are the only games around, one only needs to play naively:

Theorem 5 *Let $\{G_i\}$ each be of the form $\{x \mid 0^n\}$ or $\{0^n \mid -x\}$, no two summing to zero. If Left has a win on the $\sum G_i$, then she can win by playing on the game indicated by total ordering the incentives in Figure 4.5 by their height on the chart.*

David Moews has extended this theorem to allow additional summands of the form $\{0^n| - x\}^k$ as defined in Section 4.11. His proof, which includes the more general case, appears in [Moe].

If any two do add to zero, the Theorem is still useful, since Left can omit those two games from the sum and play on the remainder as suggested. Moews' proof of this Theorem involves characterizing all sums of winning positions, and verifying by induction that the moves suggested by the Theorem are good enough. Figure 4.5 then becomes an easy Corollary of his proof.

We can also give a good approximation to $-_x|0^n$:

Theorem 6 $-_x|0^n = n.(\downarrow\ast) - \epsilon$, *where $\epsilon > 0$ is an infinitesimal such that $m \cdot \epsilon < \uparrow$ for all integers $m \geq 0$.*

Proof: Let $g = -_x|0^n + n(\uparrow\ast)$. It is sufficient to show $\downarrow < m.g < 0$ for all $m > 0$. We'll argue about playing the games, $m.g$ and $m.g - \downarrow$, simultaneously. According to Figure 4.5, dominant moves on g for Left are on the game $-_x|0^n$ to $-_x|0^{n-1}$. Left will continue to play this term until it reaches $-_x$ and finally to $\{x|0\}$. Dominant moves for Right move $\uparrow\ast$ to 0, unless an $\{x|0\}$ is around, in which case Right will move it to 0. In fact, even if $-\downarrow = \uparrow$ is sitting around, both players will ignore it. Eventually, after the same number of moves have been made by both players, all but one of the g's are moved to 0, one is moved to $-_x$, and all the $\uparrow\ast$ terms are moved to 0. Since $\downarrow < -_x < 0$, we are done.   ∎

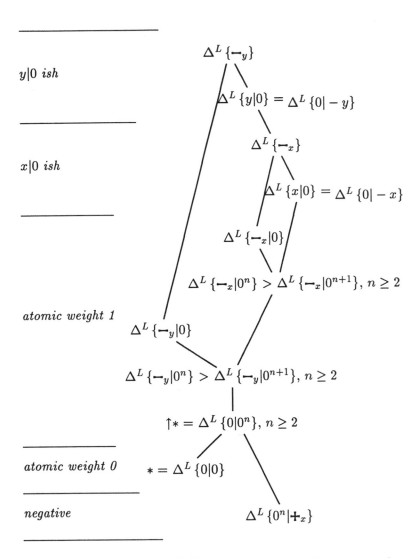

Figure 4.5: *Partial order of Left's Incentives.* x *and* y *are numbers, with* $y > x > 0$.

4.5   Rooms

Figures 4.6, 4.7, and 4.8 are catalogs of intrusions into small territories occurring in the middle of the board. The catalog is useful for seeing what hotter games naturally come up. Many of these positions can easily occur as sub-positions in one-point endgames.

Many of the positions have simple one-point kos in their game graphs. A one-point ko for Black (or White) is denoted Ⓚ (or $\overline{Ⓚ}$), and looks like:

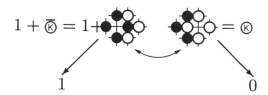

In the catalogs, below each position is its chilled and marked value in a simplified form. When it occurs, the Ⓚ symbol indicates a position which warms to a one-point ko, Ⓚ, for Black. A $\overline{Ⓚ}$ symbol warms to a one-point ko for White, $\overline{Ⓚ}$. Another new type of value, which occurs in three rooms, are games of the form G^n defined in Section 4.11.

The grid of numbers associated with each position encodes information which is useful for looking up a particular board position. Consider the subgraph induced by the empty nodes in a position. Then the l'th row of the grid and d'th number from the right indicate the number of nodes at distance l from the invading white stone and with degree d in the induced subgraph. For example, consider one of the 7-node rooms:

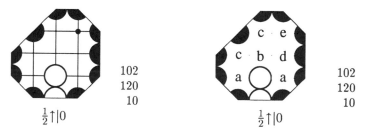

The first line of the grid indicates that of the nodes at distance 1, there are two of degree 1 (marked a), and one of degree 3 (b). The second line indicates that of the nodes at distance 2, there are two of degree 2 (c) and one of degree 3 (d). The last row means there's one node of degree 2 (e).

The rooms are then sorted lexicographically by their grid numbers. Note that different positions can have the same grid.

Again, below the room is the chilled marked value (in this example $\frac{1}{2}\uparrow|0$). Since there is one black marking in the position, the unmarked value is either $1 \int\{\frac{1}{2}\uparrow|0\}$ or $1* \int\{\frac{1}{2}\uparrow|0\}$. In this case, the parity is odd (since it's a 7-node room), and 1 is odd, so the value is $1 \int\{\frac{1}{2}\uparrow|0\}$.

One of the 7-node rooms (third row, rightmost column in Figure 4.8) has a much more complicated canonical form than the rest. Three kos are involved in its game graph (Figures 4.9 and 4.10).

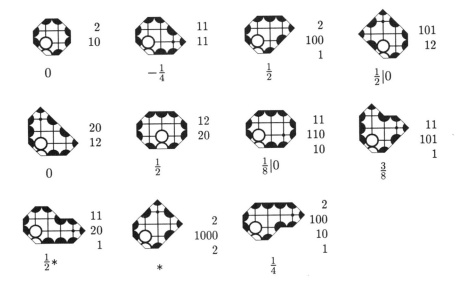

Figure 4.6: *3, 4 and 5 node rooms*

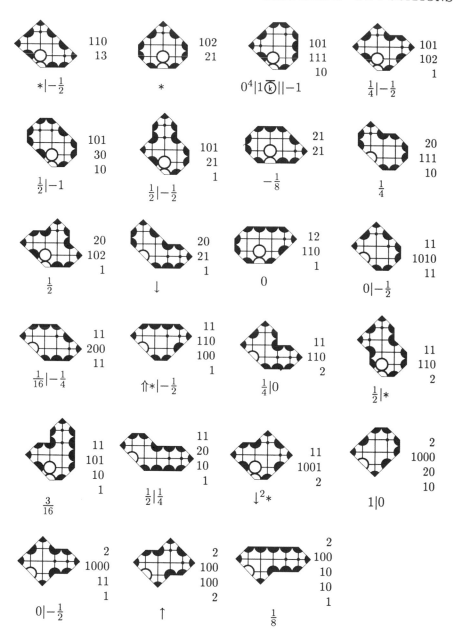

Figure 4.7: *6 node rooms*

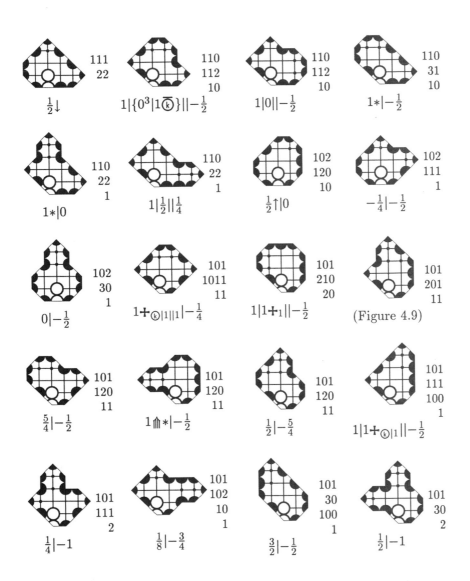

Figure 4.8: 7-node rooms

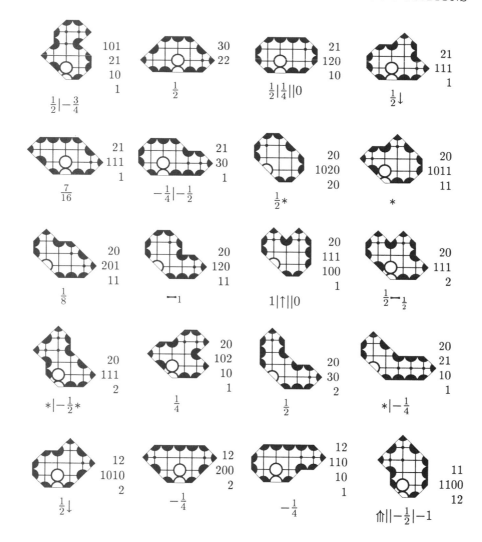

Figure 4.8: (continued from prior page)

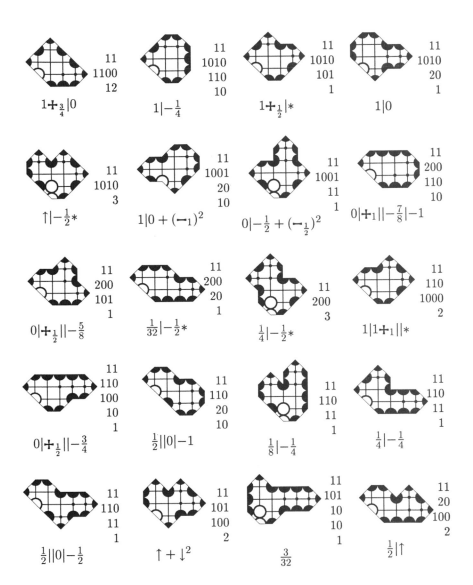

Figure 4.8: (continued from prior page)

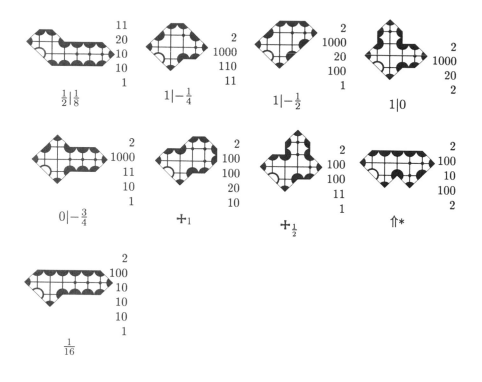

Figure 4.8: (continued from prior page)

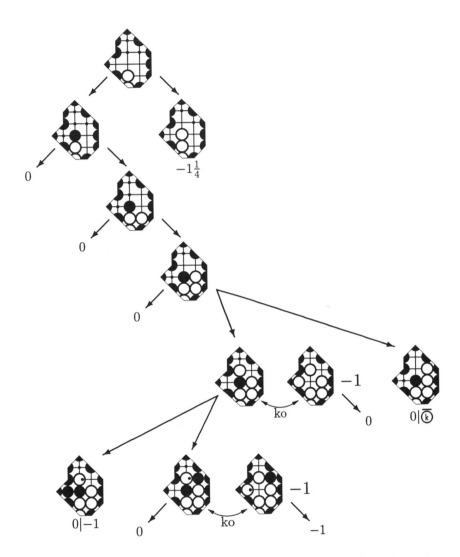

Figure 4.9: *Rogue 7-node room — the chilled game graph is shown here*

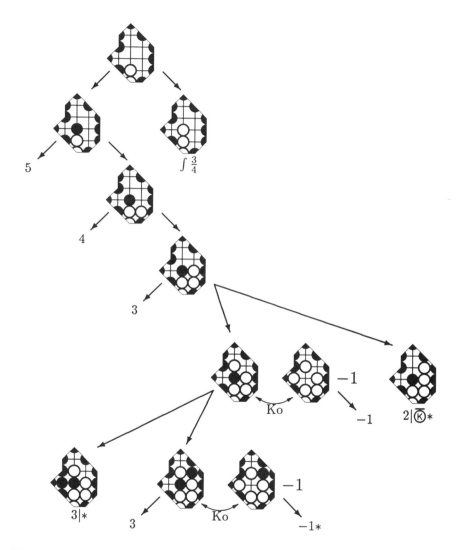

Figure 4.10: *Rogue 7-node room — the warmed game graph is shown here*

4.6 Proofs

Before giving the next class of positions, we'll want a more general technique for verifying a conjectured formula for evaluating game positions, $f(G)$. (This is not the same f used in the proof of Theorem 1 on page 53. This f simply represents a conjectured formula.)

Theorem 7 *Let F and G be games, with F in canonical form. Then $F = G$ if and only if the following four conditions hold:*

$$\forall F^L \exists G^L \text{ s.t. } G^L \geq F^L \qquad\qquad \forall G^L, \ G^L \not\geq F$$
$$\forall F^R \exists G^R \text{ s.t. } G^R \leq F^R \qquad\qquad \forall G^R, \ G^R \not\leq F$$

The conditions on the left verify that moves exist that are at least as good as those conjectured, and the conditions on the right verify that no move is *too* good.

Proof: $F = G$ if and only if the second player has a win on $F - G$, so consider playing this difference game. Since F is in canonical form, no move is reversible, so winning responses to moves on F must be on G, implying the conditions on the left. Further, no move on G can be a winning move, implying the conditions on the right. Conversely, the conditions imply the existence of winning responses on the difference game. ∎

Suppose $f(G)$ is a conjectured formula for game positions, G. We'll define the *purported left incentives*, $\hat{\Delta}^L(G)$, to be the set of $f(G^L) - f(G)$ (i.e., $\{H - f(G) : H \in f(G^L)\}$). Similarly, the *purported right incentives*, $\hat{\Delta}^R(G) = f(G) - f(G^R)$. Proofs that $f(G) = G$, will typically proceed by induction. Assume, by induction, $f(G^L) = G^L$ and $f(G^R) = G^R$. We'll then show that for each incentive there exists a dominating purported incentive, and each purported incentive is not greater or equal to zero; i.e., for Left and Right,

$$\hat{\Delta} \geq \Delta \qquad \text{and} \qquad \hat{\Delta} \not\geq 0$$

Theorem 7 above then gives us the result. When arguing on the chilled versions of games, we'll use a lower case δ instead, and show

$$\hat{\delta} \geq \delta \qquad \text{and} \qquad \hat{\delta} \not\geq 0$$

4.7 Group invading many corridors

Now, we wish to look at *unconnected* groups, which invade several corridors, as in the game G in Figure 4.11. (The lower case g in the figure is a reminder that we will be dealing with the chilled and normalized game.)

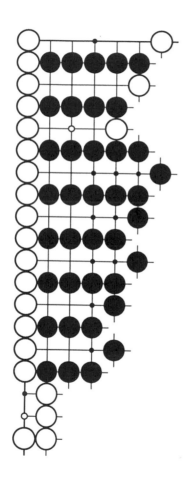

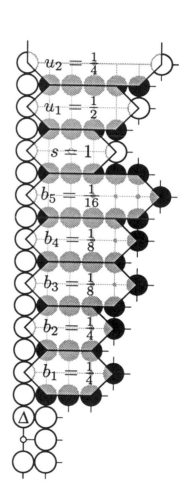

Figure 4.11:   *Game g —*
Unconnected group invading
many corridors

Figure 4.12: *Game g' — The*
same group connected at Ⓐ

Here, once Black has blocked all of the corridors on the left side, Black is threatening to capture the White group. To save the group, White must play at the node labeled Ⓐ in Figure 4.12 before Black does. We'll call this node a *socket* for the group.

The separate corridors being attacked are nearly independent, and are related only by the threat to capture the invading group. This limited interaction is just enough to complicate the situation dramatically, but not so much as to make the analyses intractable.

In order to investigate these invasions, we first use the results in Section 4.3 to see what happens if the group were connected (Figure 4.12). Call this new game g'. We define

$b_i \overset{\text{def}}{=}$ chilled and normalized values of *blocked* corridors

$\#b \overset{\text{def}}{=}$ the number of b_i's

$s \overset{\text{def}}{=}$ chilled and normalized value of *shortest unblocked* corridor

$u_j \overset{\text{def}}{=}$ chilled and normalized values of other *unblocked* corridors

If there are no unblocked corridors, we'll say $s \overset{\text{def}}{=} 0$, and if there is more than one shortest corridor, only one will be omitted from the u_j's. So now,

$$g' = \int s + \sum b_i + \sum u_j$$

We'll mark Figure 4.11 almost identically to Figure 4.12, except the socket is marked as a point for Black. Marking the socket black is convenient, since in many positions the existence of the socket will lead to a forcing play for Black, earning her about one more point. Now, to obtain the value of the original game, g

Theorem 8

$$g = \begin{cases} -1 + s + \sum b_i + \sum u_j & \text{if } \sum b_i \geq 1 \quad \text{(case 1)} \\ (1 + \sum b_i)\frac{s}{2} + \sum u_j & \text{if } \sum b_i \leq 1, \text{ and } s > 0 \quad \text{(case 2)} \\ \{x{-}2 \mid 0^{1+\#b}\} & \text{if } \sum b_i < 1, \text{ and } s = 0 \quad \text{(case 3)} \end{cases}$$

where x depends on the value of capturing the invading group.

If, as in Figure 4.11, White has exactly one space adjacent to the unconnected gap (the space just below the labeled stone in Figure 4.12), then x is

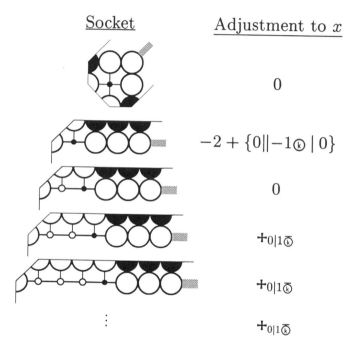

Figure 4.13: *Examples of socket environments. The shaded thick line on each diagram indicates a group of stones of any size which must eventually connect at the socket to survive.*

twice the number of stones in the invading group. Some slight adjustments to the value of x for other socket environments are shown in Figure 4.13.

In the example in Figure 4.11, $\sum b_i = \frac{13}{16}$, and $s = 1$. Therefore, we're in the second case of Theorem 8, and

$$g = \left(1 + \frac{13}{16}\right) \cdot \frac{1}{2} + \frac{3}{4} = \frac{53}{32}$$

The value of the game before chilling and normalizing is

$$G = 10 + \int \frac{53}{32} = 11* + \int \frac{21}{32}$$

Proof: The proof of the Theorem is by induction using the technique given in Section 4.6.

First, note that most moves adjust the markings by one in the mover's favor. The exceptions (which we'll argue about separately in the proof) are

- Any move on a blocked corridor of length one ($b_i = 1$)

- A White move on an unblocked corridor of length two ($u_i = 2$)

- A move by Black on the socket, capturing the invading group

Also note that a property of the formula is that dominant moves on blocked (or unblocked) corridors are on the longest one.

Base case: We'll check the formula for the few extreme cases. Recall Lemma 1.2 gives the value of a game when the left stop equals the right stop. Hence, if the socket configuration is as in Figure 4.11, the position after Black captures the white group is exactly $x - 1$, where x is twice the number of stones in the invading group. (We will not verify the adjustments to x in Figure 4.13.) So, when $\#b = s = 0$, $G = \{(x-1)|-1\} = \int\{(x-2)|0\}$ and $n = 0$. Also, by the Lemma, when there is a single blocked corridor of length one, $G = -1$ and $n = -1$. If there is a single unblocked corridor of length two, $G = -1*$ and $n = -2$. Lastly, if there are more blocked corridors of length one, each one adds $*$ to G and subtracts 1 from n. We'll assume from now on we are not in one of these positions.

Induction step: We'll show there exists dominating purported incentives. Again, as in Section 4.6, we'll use $\hat{\Delta}$ for purported incentives on G, and

$\hat{\delta}$ for purported incentives of the chilled game, g. In the latter case, moves will adjust the markings properly, one in the mover's favor.

We'll also need to show that $\hat{\Delta} \not\geq 0$ or $\hat{\delta} \not\geq 0$. When $\sum b_i \geq 1$, case 1 is greater than or equal to cases 2 and 3. When $\sum b_i < 1$, the inequality is reversed. Hence, when proving $\hat{\delta} \not\geq 0$, it is safe to ignore the possibility that a move by Black from case 1 might change cases; if the move does change cases, it only lessens the incentive. Similarly for a move by White in case 2 or case 3 when the move might change the position to case 1.

case 1: If g is an integer and it's White's move, he can connect the socket for a move with $\hat{\delta}^R = * = 1 + \int -1$. Otherwise, by the arguments in Theorems 2 and 3, when a move on the smallest valued corridor (b_i or u_j) is to a position remaining in case 1, the move has sufficiently large $\hat{\delta}$ and $\hat{\delta} \not\geq 0$. The only moves on smallest valued corridors leaving case 1 are moves by Black on the $\sum b_i = 1$. If there are u_j's, Black can play either end of those, and if there are none then Black's move is to case 3, in which case $\hat{\delta}^R$ is a negative infinitesimal.

case 2: When $\sum b_i = 1$ and $s \neq 0$, case 1 equals case 2. Assume $\sum b_i < 1$, both players move on the $b_i s/2$, $s/2$ or u_j of smallest value. Verifying $\hat{\delta}$ is ok is similar to case 1. We never have to worry about the shortest corridor changing, since in the boundary case when $u_i = s$, the formula considers the cases interchangeable. The only moves which change cases are those on positions with one corridor, the unblocked s. Here, the extreme case when $s = 2$ has already been checked, and otherwise, $\hat{\delta}^L = \{x - 2|0\}$, and $\hat{\delta}^R$ is $-s/2$ whenever $s \leq 1$.

case 3: Moves by Black are canonical. Moves by White on the longest blocked corridor either result in the $\sum b_i = 1$, or the $\sum b_i < 1$. In the former case, $g^R = 0$ and the move is canonical. In the latter, White's move is to $g^R = \{x|0^{n_b+1}\} > g > 0$. ∎

4.8 Another problem

Theorem 8 can be used to solve the following problem in Figure 4.14. Figure 4.15 shows the same problem marked and chilled.

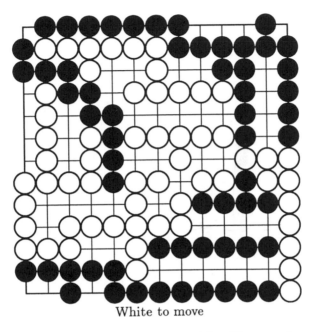

White to move

Figure 4.14: *A problem requiring Theorem 8*

There are two invading groups, one black in the upper left, and one white in the lower right. The one in the upper left is connected to blocked corridors of values $1/2$, $1/4$, $1/4$, $1/8$, and $1/16$ whose sum exceeds one, and an unblocked corridor of value $1/4$. We're therefore in case 1 of the Theorem, and all blocked and unblocked corridors are worth their value, and a -1 is added to the total. Note that this sum is negated, since it is a black group invading white, opposite from the statement of the Theorem. So case 1 gives $-(-1 + 23/16) = -7/16$.

The invading white group has a single blocked corridor of value $1/8$, and unblocked corridors of values $1/2$ and $1/8$. Therefore we're in case 2 of the Theorem. The value is therefore $(1 + 1/8)1/4 + 1/8 = 13/32$.

So, the chilled position has total value $-1/32$, and the value of the unchilled game is $\int -1/32$, perhaps plus a $*$. Therefore, White can win moving

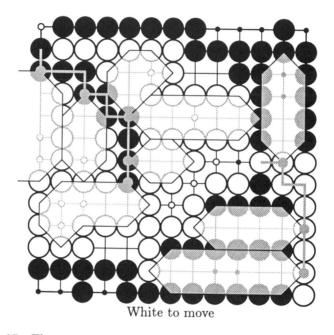

White to move

Figure 4.15: *The same problem; there are 12 black markings and 12 white* first.

The solution is summarized in Figure 4.16, which is intended to aid the player in finding the best move. Each blocked and unblocked corridor for the upper right group is labeled with its value, mirroring case 1 of the Theorem. However, the "−1" in case 1 is not indicated in the diagram. Although it's important for knowing in advance who will win, integer shifts don't affect the best moves.

Referring to case 2 of the Theorem,

$$\left(1 + \sum b_i\right)\frac{s}{2} = s/2 + \sum b_i s/2$$

Consequently, in the lower right corner, the shortest unblocked corridor is labeled with $s/2 = 1/4$, and the blocked corridor is labeled $bs/2 = 1/32$. This labeling is convenient, since it highlights the plays of greatest incentive (i.e., the corridor of value $1/32$).

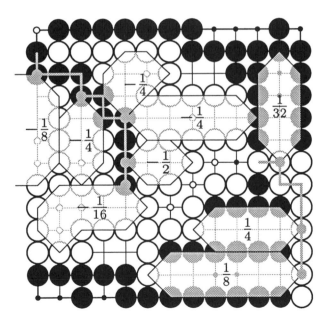

Figure 4.16: *Solution to problem*

4.9   9-dan stumping problem

The next problem we solve is the first problem given in Section 1.3. This problem is sufficiently challenging to stump every professional player who has tried it, including several 9-dan's from Japan and China.[3] Figure 4.17, along with the table in Figure 4.18 summarize the analysis. A more condensed form of the solution is shown in Figure 4.19. (As in the last Section, the Figure fails to maintain the integer part of the game values, but incentives, and therefore optimal play, can be found.)

In some cases it is not clear at first that the theorems apply; for instance, one must verify the involvement of the corner near N is of no consequence to the canonical form. So, this problem also demonstrates the robustness of the results.

By Japanese scoring, the score is $0 + \int g$, where g is infinitesimal and has uppitiness zero, but g is negative, and White can win if he plays properly. The incentives chart in Figure 4.5 aids in finding the dominant moves, the first one for White being at C.

This problem is equally challenging if Chinese scoring is used (where occupied plus surrounded territory is scored, rather than surrounded territory plus prisoners). If White wins by one point by Japanese scoring, the number of points on the board after dame are filled will be odd, and so White will fill the last dame. Therefore, White will also win by Chinese scoring. If, however, White ties by Japanese scoring, the score must still end up odd, so since Black made the first move of the game, Black will get the last dame and win. More discussion of why the mathematics usually applies to *both* Chinese and Japanese scoring may be found in Appendices A and B, especially Section B.3.7.

---

[3]In the Go rating system, 9-dan is the highest possible ranking. The total number of active 9-dans throughout the world today is about one hundred.

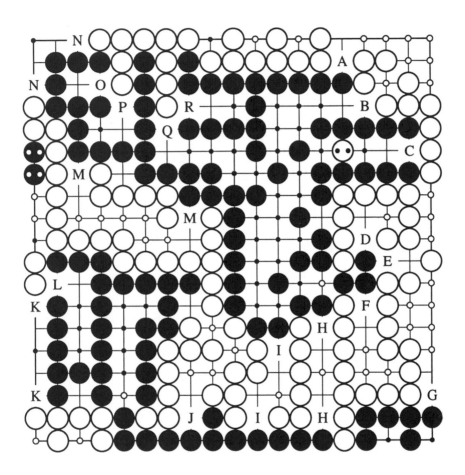

Figure 4.17: *White to move and win. There are 45 markings of each color.*

	calculations	g	uppitiness
A		$-_2|0^3$	-3
B		$0^2|+_{3/2}$	2
C		$0|+_1$	1
D	$b_1 = 1/4$		
E	$b_2 = 1/2$		
F	$b_3 = 1/16$		
	$\sum b_i < 1, s = 0$		
DEF		$0^2|+_4$	2
G		$-_2|0^2$	-2
J	$b_1 = 1/8$		
	$\sum b_i < 1$		
H	$s = 1/4$	$-1/8$	
I	$s_1 = 1/4$	$-1/4$	
J	$-1/8 \times 1/8$	$-1/64$	
L	$b_1 = 1/16$		
	$\sum b_i < 1$		
K	$s = 1/4$	$1/8$	
L	$1/16 \times 1/8$	$1/128$	
M		$-1/128$	
O	$b_1 = 1/2$		
P	$b_2 = 1/4$		
Q	$b_3 = 1/64$		
R	$b_4 = 1/4$		
	$\sum b_i = 65/64 > 1$	-1	
OPQR		$65/64$	
N	$s = 1/4$	$1/4$	
MARKINGS		0	
TOTAL		$0\pm$ smalls	0

Figure 4.18: *Calculations done to solve problem*

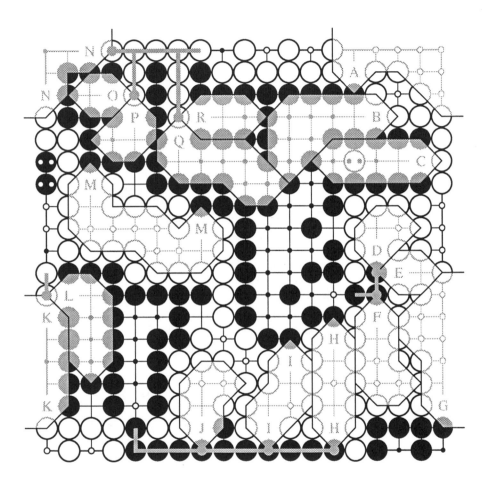

Figure 4.18: (continued from prior page)

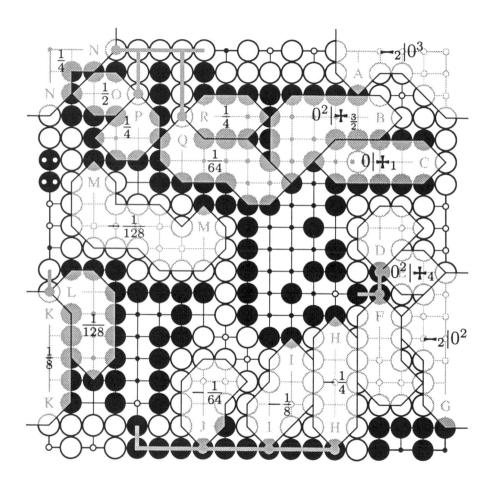

Figure 4.19: *Solution to problem*

4.10 Multiple sockets

This section gives a generalization of Theorem 8 to a tree of multiple groups with multiple sockets. An example is shown in Figure 4.20.

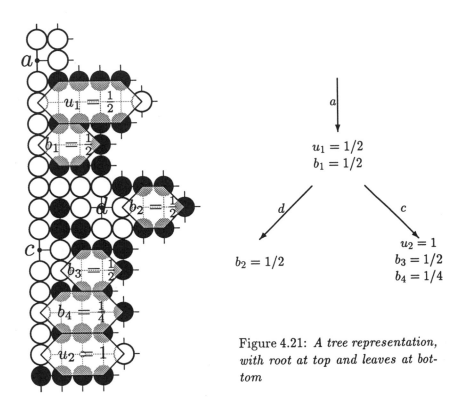

Figure 4.21: *A tree representation, with root at top and leaves at bottom*

Figure 4.20: *Multiple sockets invasion*

There are three sockets (labeled a, d and c) in Figure 4.20. The group of three white stones at the top is alive, and plays at the sockets are eventually required by White to connect the remaining groups to life. As in Theorem 8, all corridors are marked as though the sockets were filled, and the sockets are marked black.

In Figure 4.21 is a graphical *tree* representation of the same group. Each line in the diagram corresponds to a socket, indicating a play which must be made to connect a group to life at the *root* or top of the tree. Thus, to connect the group with corridors u_2, b_3 and b_4, sockets a and c must eventually be filled.

We'll assume that the value, to Black, of capturing a group is more than two points. This will be true whenever the group to be captured has more than one stone. This simplifying assumption is helpful for two reasons: (1) so that the threat to capture is worth a miny (i.e., $-_G$, for $G > r > 0$, r a number) rather than something like down, and (2) so that the threat of a small ko, after recapturing the capturing stone, doesn't become an issue.

Evaluating these multiple socket positions will be done in a more procedural fashion than positions evaluated thus far. We'll require three phases: "mate," "associate and sever," and "score."

In describing each phase, we'll use the example above. Further examples follow later, so if a detail isn't clear, don't panic!

Phase 1 (mate): Beginning at the leaves (i.e., bottom) of the tree representation, each socket will be *mated* to the largest valued unmated, unblocked corridor beneath it in the tree (i.e., in the socket's subtree). If no unmated, unblocked corridor exists, the socket is left unmated. We say the socket also has the value of the unblocked corridor, with an unmated socket having value 0.

In the example, socket d is left unmated and socket c is mated to $u_2 = 1$. Then, socket a is mated to $u_1 = 1/2$. If u_2 were unmated, then a would match with it instead, since it has a larger value and is in a's subtree. We say $a = 1/2$, $d = 0$ and $c = 1$. (Figure 4.22(a))

Phase 2 (associate and sever): Beginning with the bottom of the tree, each socket is *associated* (or, at times, reassociated) to blocked corridors' values totaling at most one in the following priority order:

1. Unassociated b-values in the socket's subtree are associated first.

2. Reassociate any b-values in the socket's subtree currently associated with larger valued sockets, "stealing" from the largest valued sockets first.

A socket associated with b-values totaling 1 is said to be *saturated*. If an unmated socket (i.e., of value 0) is not saturated, the socket and all sockets and corridors in its subtree are termed *severed*. All severed corridors can no longer be associated, reassociated or mated.

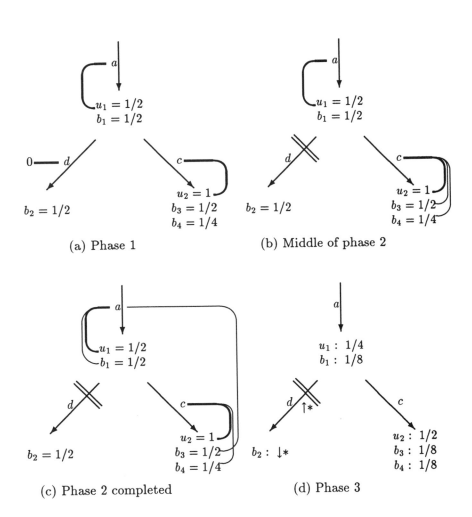

Figure 4.22: *Solution to example*

Note that the association with blocked corridor values is done in a fluid manner. A single blocked corridor's value can be divided among more than one socket when, for instance, the first socket associated gets saturated. Also, when more than one blocked corridor is available with the same priority, the formula will be consistent in all possible tie-breaks.

In the example, we begin at the bottom with sockets d and c. Socket d gets associated with $b_2 = 1/2$, which isn't enough to saturate d; since d is also unmated, it is severed. Socket c gets associated with $b_3 = 1/2$ and $b_4 = 1/4$ for a total of $3/4$ — not saturated. (Figure 4.22(b))

To complete the phase, socket a is first associated with $b_1 = 1/2$, and is still unsaturated. The only socket with larger value is $c = u_2 = 1$. So it "steals" or reassociates $1/2$ from the $3/4$ previously associated with c. (Figure 4.22(c))

Phase 3 (score): If the number of sockets in a severed subtree exceeds the number of corridors in that subtree by at least two, one of the invading white groups is dead. If the number of sockets in a severed subtree is one less than the number of corridors in that subtree, the game is *hot* since Black is threatening to capture one of the groups. (Neither of these conditions is true in the example.)

Otherwise, a contribution to the score is obtained from each of the following terms:

1. Any mated or severed unblocked corridor of value u contributes $u/2$.

2. The b-value associated with a socket of value u contributes $bu/2$. (Remember, one blocked corridor's association can be divided among several sockets.)

3. An unassociated, unsevered b-value of value b contributes b. An unmated unblocked corridor of value u contributes u.

4. Every severed socket contributes $\uparrow*$.

5. Every severed corridor contributes $\downarrow*$. (If it's unblocked, this is in addition to the $u/2$.)

Theorem 9 *Call the total contributions of 1–5 above g. If there are no severed sockets, the value of the chilled, normalized position is exactly g. Otherwise, the value of the position is $g - \epsilon$, where ϵ is a positive infinitesimal with the property that $m \cdot \epsilon < \uparrow$ for all $m \geq 0$.*

In the example,

1. Unblocked corridors u_1 and u_2 contribute half their values, or $1/4$ and $1/2$.

2. Blocked corridors b_1, b_3 and b_4 contribute $b_1u_1/2 = 1/8$, $b_3u_1/2 = 1/8$ and $b_4u_2/2 = 1/8$.

3. There are no unassociated, unsevered corridors in this example.

4. Socket d is severed, and contributes $\uparrow\!*$

5. Corridor b_2 is severed, contributing $\downarrow\!*$

So, the grand total is $h = 9/8$, and since a socket is severed, $g = h - \epsilon$. (Figure 4.22(d))

4.10.1   Consistency check

Our next exercise will be to verify that the formula is consistent with Theorem 8 in Section 4.7. In phase 1, s is mated with the socket. In phase 2, b_i's totaling at most one are associated with the socket. If $\sum b_i \geq 1$, the socket gets saturated with b-value of 1, and the remaining $\sum b_i - 1$ is left unassociated. The scoring phase yields,

$$
\begin{aligned}
g &= s/2 + 1 \cdot s/2 + \left(\sum b_i - 1\right) + \sum u_j \\
&= -1 + s + \sum b_i + \sum u_j
\end{aligned}
$$

If the $\sum b_i \leq 1$ and there is an unblocked corridor ($s > 0$), all the b-value is associated, giving case 2. Last, if $\sum b_i < 1$ and $s = 0$, the socket is severed. If the number of blocked corridors is 0, the socket is susceptible to capture; otherwise, the position has value $(\#b - 1).(\downarrow\!*) - \epsilon$, which by Theorem 6 is consistent with the formula.

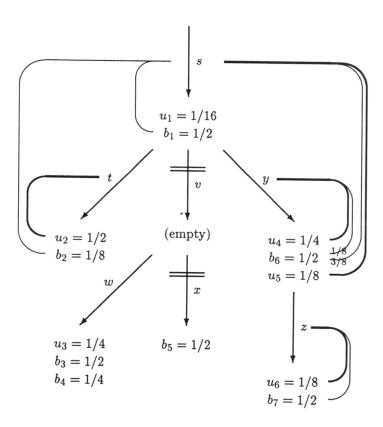

Figure 4.23: *Complicated example*

4.10.2 Another example

The example in Figure 4.23 involves many interesting cases of the Theorem.

Phase 1 (mate): First, u_2 is mated to socket t, u_3 is (temporarily) mated to w, u_6 is mated to z, u_4 is mated to y, and lastly, u_5, being the largest unmated u in s's subtree is mated to s. Sockets x and v could not be mated, as there were no unmated u's below them.

Phase 2 (associate and sever): First, temporarily associate b_2 with t.

Next, $b_5 = 1/2$ is (temporarily) associated to x, but that leaves the unmated x unsaturated, and so it is severed along with b_5. $b_3 + b_4 = 3/4$ is (temporarily)associated with w. v reassociates the $3/4$ from w, leaving v unsaturated. v cannot reassociate the $b_5 = 1/2$, since that is already severed. So v, being unsaturated and unmated, is severed, along with the whole subtree.

Now, b_7 is associated with z. b_6 is associated (temporarily) with y, leaving it unsaturated at $1/2$. (It can't reassociate from the smaller valued $z = 1/8$.)

Last, $s = 1/8$ associates with the only unassociated, unsevered b available in its subtree, i.e., $b_1 = 1/2$. Being unsaturated, it steals $b_2 = 1/8$ from t, the largest valued socket in its subtree. It now needs another $3/8$ to be saturated, and so it reassociates that $3/8$ from the next larger valued socket in its subtree, b_6. b_6's association is now split, $3/8$ to s, and $1/8$ to y. (This brings us up to date with Figure 4.23.)

Phase 3 (score): We notice the socket values are $s = 1/8$, $t = 1/2$, $y = 1/4$, and $z = 1/8$

1. u_2 contributes $1/4$, u_3 $1/8$, u_4 $1/8$, u_5 $1/16$, and u_6 $1/16$.
2. b_1 contributes $1/32$, b_2 $1/128$, b_6 $3/128 + 1/64$, and b_7 $1/32$.
3. u_1 contributes $1/16$.
4. v, w and x each contribute $\uparrow*$.
5. b_3, u_3, b_4 and b_5 each contribute $\downarrow*$.

So the game is worth $51/64 + \downarrow* - \epsilon$.

Notice in the example how b_2 multiples $u_5/2$, but in fact that a move on b_2 has chilled incentive $-1/64$ (!)

4.10.3 A "realistic" example

Although the position is contrived, the following example demonstrates how multiple sockets might look in a game.

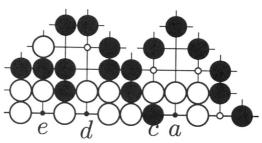

Recall that very short White corridors can have Black markings as in Section 4.3. Here, a, c, d and e are all sockets. (Black's play at a captures White's stone at c, and then Black can play at c to connect her group.) The stone at c is not marked since the effects of being both a captured stone (marked white) and a socket (marked black) cancel. There is one unblocked corridor of value 1, and blocked corridors of values 1, 1, 1, and 1/2. One of the sockets mates with the $u = 1$, and the other three are unmated. The blocked corridors of value 1 associate with the unmated sockets, saturating them, and the last $b = 1/2$ associates with socket of value $u = 1$. The value of the position is $u/2 + bu/2 = 3/4$.

4.10.4 The economist's view

Another way to formulate the multiple sockets theorem is to think of the blocked corridors as *consumers* and the sockets as *suppliers*. Each supplier has unit *capacity* (the amount it can produce), and each consumer has demand b corresponding to its b-value. For convenience, each u-value will be replaced by $2h$, so $h = u/2$. Each supplier's h-value will represent its *cost* of production of one unit.

Keeping in mind the economic analogy of suppliers and consumers, the object is to set up distribution contracts which supply the consumers. The consumers try to obtain the minimum price possible, and the competing suppliers try to maximize profit, i.e., price minus cost. Each supplier is able to distribute only to consumers in its subtree and is required to charge the same price to all its consumers.

Again, we'll describe the algorithm in phases. The first phase, "mate," is the identical to the first formulation of the Theorem in Section 4.10.

Phase 1 (mate): Beginning at the leaves (bottom) of the tree representation, each socket will be *mated* to the largest valued unmated, unblocked corridor beneath it in the tree (i.e., in the socket's subtree). If no unmated, unblocked corridor exists, the socket is left unmated. We say the socket also has the value of the unblocked corridor, with an unmated socket having value 0.

Phase 2 (initialize): We assign an initial, feasible set of distribution contracts by fulfilling each consumer's requirements from its own *local supplier* at unit cost. These local suppliers have no lasting effect, and are introduced only to streamline the text.

Before describing the bidding phase, we'll require a few definitions. A supplier is *profitable* if its price exceeds its cost. It is *saturated* if it is contracted to supply one unit. A supplier is *idle* if it has no distribution contracts. If a supplier is neither saturated nor idle it is *unsaturated*.

Phase 3 (bid): Beginning at the leaves (bottom of the tree), each socket tries to steal contracts from sockets in its subtree by making competitively priced *bids*. Initially the bid price is one, and all suppliers in the current subtree which are charging the bid price are said to be *competitors*. Other suppliers may have lower prices, but any suppliers in the current subtree with higher prices are idle.

If the supplier at the root of the subtree is profitable and unsaturated, then the supplier steals contracts from any unprofitable competitors. If there are no unprofitable competitors, then the bid price is lowered until either it matches the price of new competitors or the supplier at the root becomes unprofitable.

If the bid price reaches zero (i.e., the root can get no contracts), then the subtree is severed as in phase 2 of the first formulation in Section 4.10. If the supplier at the root is unprofitable or saturated, then continue on to the next supplier — the bidding round for the current supplier is over. Otherwise, the supplier continues to lower its bid.

Observe that after the bidding phase, the following hold true:

1. Although contract prices may differ, each supplier charges a single price to all its consumers, and each consumer pays a single price to all its suppliers.

2. All profitable suppliers are saturated.

For each contract, the profit is the difference between price and cost. Let $P(b)$ be the price charged to consumer b and $P(h)$ be the price charged by supplier h. The total price can be measured from the point of view of the consumers, b; the total profit can be measured at the saturated suppliers, h, since all profitable suppliers are saturated:

$$\text{COST} = \text{PRICE} - \text{PROFIT}$$
$$\text{PRICE} = \sum_b b \cdot P(b)$$
$$\text{PROFIT} = \sum_{h \text{ satur}} P(h) - h$$

Theorem 10 *The value of the chilled normalized multiple sockets position is infinitesimally close to*

$$\text{COST} + \sum_h h$$

As in Theorem 9, there may be lower order terms due to severed sockets.

4.11   Infinitesimals generalizing up-second

In Section 4.10, when subtrees were severed, results were bounded to within an error term that was a higher order infinitesimal than ↑. In this section we precisely evaluate this higher order infinitesimal for some cases. These should indicate some of the difficulty in completely characterizing all multiple socket positions. The results are given without proof.

First, we define some new infinitesimals which are a generalization of $\uparrow^n$ (pronounced "up nth") [BCG82, p. 227] suggested by Moulton [Mou90]. For $G = 0|H$ infinitesimal in canonical form, define:

$$G^0 = -H$$
$$G^i = 0 \left| \left(-G^0 - G^1 - G^2 - \ldots - G^{i-1} \right) \right.$$

$$G^{\to 0} = 0$$
$$G^{\to i} = G^{\to i-1} \left| H \right.$$
$$= G^1 + G^2 + \ldots + G^i$$

Also, negating G in the above definitions negates the right hand side.

For $G > 0$ infinitesimal and $i > 0$, G^{i+1} is infinitesimal with respect to G^i:

$$\text{for all } n > 0, \ 0 < n.G^{i+1} < G^i$$

Using the same notation as in Section 4.10, we'll look at the class of positions shown in Figure 4.24

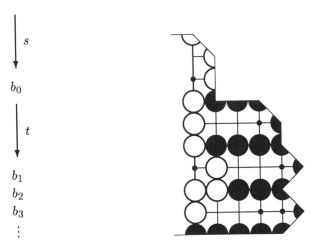

Figure 4.24: *Position with interesting infinitesimals. Shown on the right is an example where $b_0 = 1/4$, $b_1 = 1/4$ and $b_2 = 1/8$.*

Suppose the group t has X stones, and suppose $x \approx 2X - 2$ is the chilled value of capturing the group. Let $\#b$ be the number of blocked corridors below t. Further let i be the length of the blocked corridor b_0. The value of the chilled, normalized position, g, is given by:

1. If $b_0 = 1$, then g is identical to the socket position beneath t.

2. If $\#b = 0$, then $g = \{x|0\} + {\longrightarrow_x}^{i-1}$

3. If $\displaystyle\sum_{i \geq 1} b_i \geq 1$ and $b_0 + \displaystyle\sum_{i \geq 1} b_i \geq 2$, then $g = \displaystyle\sum_{i \geq 0} b_i$.

4. Otherwise, $g = \left\{ -_x | 0^{\#b-1} \right\}^{\to m}$, where m is the number of moves on b_0 before the previous condition 3 occurs.

What's surprising about the positions with minies, is that the move on b_0 remains in canonical form for White.

Lastly, we consider the Go position in Figure 4.25. By applying Theorem 9, we can conclude that the value is $\downarrow* + \epsilon$. In this case, the higher order infinitesimal, ϵ, is quite complicated. In fact, the Go position has canonical form:

$$\Big\{ 4\{-_8|0\} , \{4|0\}-_4 \,\|\, -_{16}|0\Big\}, -_4 \,\Big|\, 0 \approx \downarrow*$$

The existence of the $-_4$, $-_8$ and $-_{16}$ in the canonical form is a good indication it does not break up into much simpler sums.

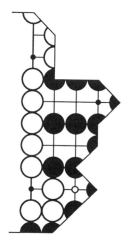

Figure 4.25: *A surprisingly complicated position*

Chapter 5

Further Research

The most natural directions to extend the theory are to include other parts of the game, with the hope of developing new mathematics, or improving Go playing programs, or both.

5.1 Applying the theory earlier in the game

The results in this book apply most directly to the late-stage endgames. However, the techniques also simplify the analyses of earlier endgame positions. For instance, the game trees for the rooms given in Section 4.5 were substantially simplified by chilling, even though most of the games had temperature exceeding one. Also, the strongest Go players often make moves based on small endgame subtleties long before the endgame arises.

5.2 Approximate results

In some sense, the biggest problem currently facing the theory is also its biggest virtue, that being its emphasis on being exact. The precision that is gained is sometimes at the cost of generality.

We give useful approximate results in Theorem 9, bounding values of some games to within an amount infinitesimally smaller than $\uparrow$. But, there are other ways one could try to give approximate results.

One possibility is to find conditions under which cooling by more than a point is useful. The cooling might preserve enough of the structure of some game trees that the mover can use the cooled game to find nearly optimal plays.

There are several obvious impediments to progress in this direction. Although one can reheat positions that have been cooled by more than one degree, this mapping is usually not one-to-one. Further, the best move is not always in the hottest game, so cooling can hide the very best move. Further discussion of this topic appears in Section 5.7.

A second approach toward approximate results is to extend the theory to account for partial information. One way to model the partial information is by randomness. The values of a subtree could be randomly distributed, reflecting the uncertainty of the player or computer program about the outcome. How should a player play on a sum of random games so as to maximize his probability of winning? There are two problems to overcome in this approach:

- As soon as a game has random outcomes, the game plus its negative is no longer a win for the second player. So all of the group structure is gone.

- It appears that few games would be comparable in any reasonable partial ordering of games, which lessens the simplifications occurring in conversion to canonical form.

Consequently, major changes to the theory would probably be required.

Another approach is to give bounds on the value of a game when bounds on subgames are known. However, small changes to a subgame can lead to larger changes in the original game. For instance, $\{0|2\} = 1$, while $\{\downarrow|2\} = 0$, so an infinitesimal adjustment to a subgame can explode into an integer change in the original game.

5.3 Kos

The introduction of one-point kos (denoted $\circledK$ as in Section 4.5) does not detract from the results substantially. The Japanese value may be bounded by

$$* \leq \circledK \leq \{1*|0\}$$

The lower bound is obtained by assuming Left can never win the Ko, and the upper bound is when Left can always win the Ko. So in the chilled game,

$$0 \leq \circledk \leq 1/2$$

Also, by using the ban on repeated positions, there is a sense in which

$$Ⓚ - Ⓚ = 0 \quad \text{and} \quad Ⓚ + Ⓚ + Ⓚ = 1$$

The incentive to move on Ⓚ is at most $\int -1/2$, and a consequence of Lemma 1.5 is that the only moves with worse incentive are on unheated $*$ (i.e., dame). Consequently, kos will not be played until the end of the game, and their value is limited.

Optimal play for Left in an elementary Go game with one-point kos is to maximize, in this order,

1. points in the ordinary warmed values

2. one-point kos

3. kothreats

If Black has only one move which maintains a fixed number of points in the first term, it is optimal.

The same cannot be said in non-Japanese rules. In Chinese rules, the winner of the ko may earn two points, and if there are no dame, the loser of the ko may be forced to just fill territory or pass.

Larger and more complex kos require more modifications of the theory. The concept of a sum of games breaks down somewhat, since legality of a play on one summand is determined by more global considerations.

5.4 Life and death

Landman proposes using combinatorial game theoretic techniques for studying life and death problems of groups [Lan]. Suppose a group is susceptible to attack, and needs to make two eyes. There may be several separated regions of play to make eyes, and each region is completely analyzable. Each of these regions has its own game tree, where the score is the number of eyes made.

At a strategic level, there are only three distinguishable outcome classes: 0 eyes, 1 eye and ≥ 2 eyes. Landman approximates the ≥ 2 class by 2, and he then gives examples of positions which he models with the values 0, $\frac{1}{2}$, $\frac{3}{4}$, 1, $1*$, $1\frac{1}{4}$, $1\frac{1}{2}$, and 2 eye(s). Since all stopping positions are 0, 1 and 2, as a consequence of the proof technique in Lemma 1.5, these are the only chilled values that can occur. But, linearity does not always work in Landman's model. The fact that 3 belongs to the ≥ 2 class (which simplifies to 2) implies other bizarre simplifications, such as the simplification of $\int 1 + \int \frac{3}{4}$ to $\int 1\frac{1}{4}$.

When two groups are in *semeai* with enough shared liberties, having one eye can be important. If Black has one eye and White has none, then Black can capture White. If, however, Black and White both have one eye, or if both have no eyes, then they will both live in seki. By modeling the values of potential eyes in the shared regions between the groups, Landman has obtained a collection of partial results. As in the single group case, linearity fails, and the model bears only superficial resemblance to classical combinatorial game theory.

5.5   The last play

There are many phases of the game where getting the last play is important. It is often important to get the last big opening move, before the incentives for moving drop considerably, or the last play in the midgame before endgame begins, or the last of a collection of 8–10 point endgame moves before the values drop to 3–4. The same techniques to understand the last point should help here as well. The key concept is games with infinitesimal values.

5.6   Extensions of current results

More tractable research areas include extending the multiple corridor theorems given in Sections 4.7 and 4.10:

- Extend the catalog of positions given in Section 4.5 to include rooms bordering the edge of the board, or rooms in which more than one invading stone penetrates the territory.

- Include more complicated corridor structures. There might be "dead" stones along a corridor, or a room at the end. Some preliminary results in that direction appear on pages 196–201.

- One wall of a corridor may not be immortal (live), and Black may need to connect it with a blocking move on the corridor. This gets interesting when the corridor has bends, and blocking moves do not always connect the two walls. Typically (as in the corridor marked M

in Figure 4.19 of Section 4.7), as long as enough moves connect the two walls, the values are unaffected by assuming immortality.

- Allow the opposite end of an unblocked corridor to be another (multiple) socket invasion rather than a live stone.

- An unblocked corridor with bends may have both ends invaded by the same group.

- Give more precision than the $\uparrow*$'s in Theorem 9 of Section 4.10. In particular, perhaps one could include lower order terms of the form $\uparrow^n$.

- Rephrase Theorem 9 so that the associated blocked corridor which has highest incentive can be identified by its contribution in the scoring phase.

5.7 Hardness results

Lichtenstein and Sipser have constructed Go positions without kos which are PSPACE-hard [LS80]. Using kos, Robson has shown Go is EXPTIME-complete [Rob83]. Both of these results involve constructing large life and death situations taking over much of a very large board.

Lockwood Morris [BCG82, p. 219] first showed that playing sums of relatively simple combinatorial games is NP-hard. Yedwab and Moews subsequently showed that even a sum of games of the form

$$a\|b\mid c$$

is NP-hard. Since, once heated, any such game can be readily constructed on a sufficiently large Go board, we think this result is of sufficient interest that we will restate it herewith.

Theorem 11 *The following problem is NP-hard:*

INSTANCE: *A collection of games, each of the form $a\|b\mid c$, where a, b and c are integers.*

QUESTION: *Can Left win moving first on the sum of all the games?*

Proof: Yedwab [Yed85, pp. 29–45] actually proved that playing sums of games is PSPACE-complete even if the individual games are each represented

by a tree of depth two with at most two options by either player. Her proof
consists of a clever reduction from a variant of the SUBSET SUM problem
— given a list of integers, can it be partitioned into two subsets of equal
sum? The particular variant used for the NP-hardness proof is:

INSTANCE: A set X of n integers, $X = \{x_1, \ldots, x_n\}$ and a target S.

QUESTION: Is there a $k \leq n$ and a $\{y_1, \ldots, y_k\} \subset X$, such that $y_1 > y_2 > \cdots y_k$ and the alternating sum

$$S = y_1 - y_2 + y_3 - \cdots \pm y_k \ ?$$

This problem is known to be NP-complete.

Yedwab constructed a set of games whose sum is negative unless and only
unless the set X and the number S have that property. One of Yedwab's
summands had multiple options; Moews [Moe] later refined the construction
to replace it with a pair of games each of which had unique options.

The sum of the games in the refined construction may be denoted as

$$\sum G_{i,1} + \sum G_{i,2} + H + I_1 + I_2 + \Uparrow \tag{5.1}$$

The temperatures of the G's, H, and I's are contrived to be sufficiently big,
and sufficiently far apart, that, for the first $2n + 1$ moves, neither player
can gain by playing anywhere other than in a hottest summand. $G_{i,1}$ and
$G_{i,2}$ will have equal temperatures, t_i, which might, for example, be taken as
$K3^{n+3-i}$, where K is suitably larger than S and every $x \in X$.

Yedwab's first $2n$ summands are

$$\begin{aligned} G_{i,1} &= t_i \parallel -t_i* \\ G_{i,2} &= t_i \parallel -t_i+x_i \mid -t_i-x_i \end{aligned}$$

Each of these games has mean zero. But the t_i are sufficiently big and
spaced sufficiently far apart that when Left plays either of the above games,
Right's only hope is to respond on the other. Thus, Left may pick any subset
of the n numbers in X, and play her first n moves in such a way that after
these moves and Right's responses, all that remains of the G's are whatever
subset of switches Left has chosen. Left must then play on the next hottest
game,

$$H = 0 \mid -2t_{n+1}*$$

and Right then gets to choose between these two games:

$$
\begin{aligned}
I_1 &= t_{n+3} \mid -S \parallel -t_{n+2} \\
I_2 &= S + t_{n+2} \mid *
\end{aligned}
$$

The temperatures of the I's are close enough that Right can seriously consider playing on either. If the alternating sum of the relevant x's exceeds the number S, then Right can win by playing on I_1. If the alternating sum of the relevant x's is less than S, then Right can win by playing on I_2. So Left wins against a competent opponent if and only if she was able to select a subset of the x's whose alternating sum is exactly equal to the number S. ∎

Although Left's margin of victory in the Yedwab/Moews construction is only ⇑ or ⇑*, the net score can easily be increased by appropriately heating all summands. However, the temperature of the original sum in equation 5.1 seems to be much larger than the net score, and there was at most one occasion in the entire game when a player moved on the second hottest summand in preference to the hottest one.

There is still a large gap between Go positions which are provably hard and those analyzed here, and there is ample room for numerous further results of both types.

Appendix A

Rules of Go — A Top-down Overview

In this Appendix, we present a top-down overview of the major rule systems. An alternative bottom-up approach is presented in Appendix B.

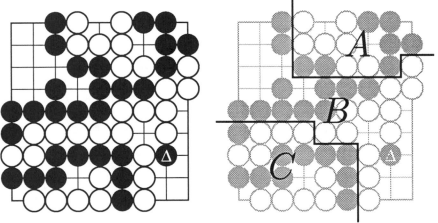

Figure A.1: *A position where all territorial scoring rules agree*

Figure A.1 shows a game in which all major rulesets agree. The marked black stone is dead. If play continues in a competent manner, the outcome should be a tie. Region *B* includes the dead stone and all of the territories surrounded by independent live groups. Region *A* contains a seki with two one-eyed groups. Region *C* contains a different, more common kind of seki, the two groups that have no eyes. All major rulesets agree that the net score within each of these three regions is 0.

A.1   Rulesets can (rarely) yield differing results

Consider next the game shown in Figure A.2. It is now Black's move. If both players play competently for the rest of the game, then what will be the final difference between Black's score and White's score? Who won? The table of net scores in Figure A.3 shows that your answer depends very much on your particular dialect of rulespeak.

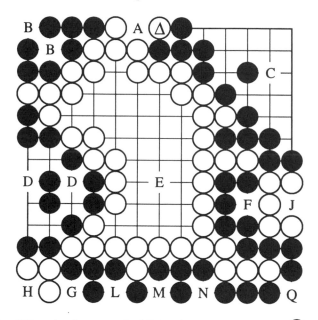

Figure A.2: *What is the score? White has just played at ⓪, capturing a single black stone in ko. Each side has 1 captive, and 52 stones on the board, none of which are dead.*

A.2   Who has used these different rulesets?

The table in Figure A.3 shows five important scoring systems. The first two are predominant in Japan and China today. The Ing rules have been used in tournaments sponsored by the Ing Foundation of Taipei. Ing is a

Region	Japan	China	Ing 1986	Ancient	Math
A	0	-1	-1	-1	$\checkmark$
B	$+2$	$+2$	$+2$	0	$\checkmark$
C	$+18$	$+18$	$+18$	$+16$	$\checkmark$
D	$+8$	$+8$	$+8$	$+6$	$\checkmark$
E	-28	-28	-28	-26	$\checkmark$
F	0	0	$\frac{2}{4}-\frac{2}{4}=0$	0	0
G	0	0	$\frac{2}{3}-\frac{1}{3}=\frac{1}{3}$	0	0
H, J	0	-2	-2	0	0
L, M, N, Q	0	$+4$	$+4$	$+4$	$+4$
TOTAL	0	$+1$	$+1\frac{1}{3}$	-1	

Figure A.3: *Score of Figure A.2 under different rules. A $\checkmark$ indicates that mathematical rules can be designed to match any of the scoring methods.*

leading sponsor of computer Go competitions, of junior world competitions, and of the Ing cup professional competition. The Ing rules are international; they differ slightly from the official Taiwan rules. The ancient rules, or some variation of them lost in antiquity, were once used in mainland China, where the game of Go originated about 4,000 or 5,000 years ago. In Section B.2.2, we show how a variation of these ancient rules can provide the foundation for a bottom-up understanding of several traditional concepts which are otherwise hard to define precisely: *eye, big eye, dead, independently alive, seki.*

The various rulesets which are most tractable for pursuing formal and rigorous proofs of the most general and useful theorems about Go endgames are based on the mathematical foundations of combinatorial game theory [BCG82] [Con76]. All such rulesets essentially prohibit passing, and transform conventional score-counting into a contest to see who can get the last legal move. Applied mathematics is partly the art of formulating models which are both relevant and tractable. In that spirit, mathematicians are challenged to formulate rules which *mathematize* each of the popular rulesets of Go as closely as possible. Our work in that direction is included in Appendix B.

The vast majority of Go endgame problems are independent of the ruleset. We have found a class of mathematical rulesets which are especially well-suited to analyzing such problems, and we call such mathematical rule-sets *Universalist*. The scores given by such dialects for the various regions of Figure A.2 are also shown in the table of Figure A.3. A $\sqrt{}$ means that one can select a subdialect of the Universalist rules whose score will coincide with any other prior entry in that row.

The Go associations of Taiwan, New Zealand, and North America have formulated their own official sets of rules. Several other interesting modern rulesets have been formulated by Ikeda Toshio [Ike92]. We conclude this appendix with our top-down comparison of some of these major rulesets, followed by brief discussions of some of their distinctive differences. The reader may find our thoughts on some additional subtleties of traditional and current Japanese rules near the end of Appendix B.

Many of the world's Go associations have active rule committees. Some of the peculiarities of some rule systems, including the 1986 Ing rule for prorating dame in seki, may already be historical curiosities by the time the reader reads this appendix. However, the pace of rule changes is measured in decades rather than in months or years. [KOM86]

A.3   Top-down view of rule options

Here

A	=	ancient
C	=	modern China
J	=	Japan
U	=	Universalist (A class of mathematical rulesets)
NA	=	North American
NZ	=	New Zealand
Ing	=	Ing 1986
T	=	Taiwan

> In most cases, when any of the above is un-mentioned in any particular rule, it defaults to agree with **C**. More thorough discussions of the rules may be found in the references.

0. Initial Conditions: *Komis and/or handicaps may vary according to the tournament. (See Section B.3.3.) The standard board size is 19 × 19, although any rectangular size could be used. Beginners often prefer 9 × 9 or 13 × 13, which lead to shorter games, and there is also some literature on Monster Go [Fai90b] played on boards of size 23 × 23 or larger. In games with no handicap, Black conventionally moves first.*

1. Capturing: *The capturing rule, as defined in Section B.2.1, is universal among all dialects of the rules which we consider in this book.*

2. Suicides:

    2a. C, J, U: *Any suicidal move is illegal*

    2b. T, Ing, NZ, U: *A suicidal move which sacrifices a single stone is illegal, but a suicidal move which sacrifices ≥ 2 stones is legal. See Figure A.4.*

3. Loopy Play:

    3a. NZ, NA, U: *A move which repeats any previous board position is illegal.*

    3b. J, U: *A move which repeats the same position of the board 2 moves ago is illegal, but a move which repeats the same position 4, 6, or more moves ago is legal. See Figure A.5.*

    3c. U(Conway): *There is no ban on moves which repeat previous board positions. See Section A.5.*

4. Passing:

    4a. U: *Instead of playing a stone on the board, a player may return one captive stone of the opponent's color to the pot.*

    4b. C: *Instead of playing a stone on the board, a player may take a stone of his color from the pot and give it to the opponent as a captive.*

    4c. J: *Instead of playing a stone on the board, a player may pass without penalty.*

5. Termination: *Play continues until either player chooses to resign* **or** *both players agree on the final score* **or**

5a. U: *One player has no legal move, in which case the game ends and he loses.*

5b. NA: *Black passes and then White's next move is also a pass.*

5c. C, J: *There are two consecutive passes, with either player passing last.*

6. **Scoring Philosophy:** *If consecutive passes terminate the game and the players fail to agree on the score, then*

   6a. C: *My score= My komi + Stones on board of my color + My territory*

   6b. NA, U: *My score= My komi + Captive stones of opponents' color + My territory*

   6c. J(traditional): *The dispute is submitted for arbitration to the authorities, whose decision, following rule 6b as a generic guideline, should take into account the history of precedents of prior rulings in other disputed games.*

7. **Aesthetic Appeal:**

   7a. C: *Ultimate Modularity. Each node on the board is assigned one point, which is awarded to a player or prorated between them. The assignment of each point is based entirely on local considerations.*

   7b. U: *Functional Modularity. An endgame decomposes into a sum of partial board positions, about which useful general theorems can be rigorously proved and usefully applied.*

   7c. J: *Games should end before either player needs to fill in his own territory. In our jargon, "No childish encores."*

   7d. NZ: *Simplicity. Rules should be concise, logically complete, and unambiguous.*

8. **Territorial Scoring:** *(See Figure A.1). Explicitly or implicitly, my territory = number of nodes surrounded by independently alive stones of my color, plus sum of adjustments indicated in all subsequent items below.*

9. **Independent Live Groups:**

   9a. A, U: *2 points for each of opponents' independent live groups*

9b. C, U: *0 points for each of opponents' independent live groups*

10. Unfilled active ko

10a. C, U: *1 point for a node surrounded by my stones in unfilled, active ko.*

10a. J, U: *0 points for a node in unfilled, active ko.*

11. Private eyes in seki

11a. C: *1 point for each node in each of my private eyes in seki*

11a. J, U: *0 net points for each of my private eyes in seki*

12. False eyes in seki

12a. C, U: *1 point for each of my false eyes in seki, or for any other empty nodes in seki where I can play but my opponent cannot play (cf. Section B.5.6).*

12a. J: *0 net points for each of my false eyes in seki*

13. Dame in seki

13a. C, U: *0 net points are given for dame in seki*

13b. Ing: *For each dame in seki, one point is prorated according to colors of stones on adjacent nodes.*

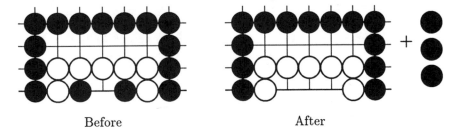

Before   After

Figure A.4: *A suicidal Taiwanese kothreat, before and after the move*

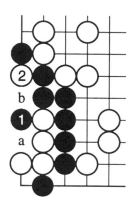

Figure A.5: *"Eternal life"* — *A Japanese loop of length 4. Best play is Black at a, White at b, Black at* ❶, *White at* ②, *returning to the same position.*

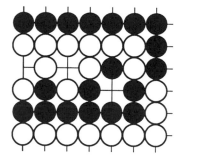 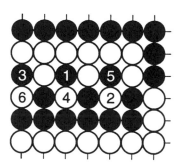

Figure A.6: *A triple ko (in seki)* — *Black to move. After the sequence shown on the right, each move capturing a single stone, the position is repeated.*

A.4 Interpretations of territories

As illustrated in Figure A.3, significant differences in outcome between the various sets of rules can result from differences in what these rules imply about how territory should be counted.

In all scoring systems, it should be permissible to remove dead stones from an enclosed territory (such as the marked stone in Figure A.1), before counting the score. The issue of how to identify which stones are dead is discussed in Section B.4.2. This is usually obvious to competent players. In any scoring system except Japanese, there is no need for the rules to identify dead stones and enclosed territory, because it costs a player nothing to refrain from passing until after he has captured all opposing stones that he claims are dead.

Japanese scoring assigns net scores *only* to territory enclosed by independent alive groups, such as region B in Figure A.1. There are no territorial points for sekis, such as regions A and C in Figure A.1. The distinction between an independently alive group and a seki rests on global considerations, which we pursue in Section B.2.8.

Chinese scoring uses a different definition of territory, which depends only on very local properties, applied only after no dead stones remain on the board. Any empty point is territory if it is part of a connected empty region which is adjacent only to one color; it then counts as territory for that color. If the perimeter of the empty area is occupied by both colors, then the points corresponding to those nodes are split between Black and White. In Chinese scoring, the split is 50%-50%. In Ing's rules, the split was prorated according to the number of stones of each color occupying the perimeter.

Perhaps the most common scoring difference arises from the unresolved ko, as in Figure A.2. In Japanese scoring, it is worth nothing, because it is certainly not surrounded by independent life. But in Chinese scoring, since it is surrounded by White, it counts as a white point. Like any other point of white territory, it makes no difference to the Chinese score if white subsequently fills it with a white stone.

A.5 Loopy play and hung outcomes

Rule systems differ in their rules concerning the circumstances, if any, under which it is permissible to repeat a prior board position, and then possibly continue to play through the same sequence of loopy positions again and again. These differences are much better known (but much rarer in relevance) than the difference in *ko scoring* rules. Western (i.e., both New Zealand and North American) rules forbid any move which repeats any prior board position, whether it occurred 2 moves ago, 4 moves ago, 6 moves ago, or more. Japanese rules forbid only 2-cycle loops. Although 4-cycle or 6-cycle Japanese loops are rather rare, they can occur. Examples are shown in Figures A.5 and A.6. More examples can be found in [Har88]. We use the English word *hung* to describe a game in which play gets stuck in such a repetitive loop. Although this matches the American computer jargon, as in *hung in a loop*, the Japanese connotation is more similar to the American notion of a *hung jury*. Although the situation is logically very similar to the notion of a draw by perpetual check in western chess, the cultural connotation is very different. Japanese authors insist that hung even games are *not* ties, but rather *no result*. At the Nihon Ki'in, a hung game is replayed later.

The Official 1988 Chinese rule on loopy positions is somewhere intermediate between the western rule 3a and the Japanese rule 3b. The Chinese rules prohibit some types of loopy positions, but authorize the referee to declare others to be hung. Section 20 of the official 1988 Chinese rules includes examples of each type. We make no effort to formulate a general algorithm that might attempt to predict how the Chinese rule authorities might adjudicate a loopy position that differs from all of their published examples.

A.6   Protocol

Officials of competitive sports must occasionally make big decisions about how to handle the behavior of players which is not directly related to the rules of the game. It is occasionally necessary to expel a player for *unsportsmanlike conduct*. Negotiations involving the conditions under which superstars agree to play can receive more coverage on the sports pages than the games themselves.

Protocol issues can also play a big role in intellectual games. In the 1970s, when Bobby Fisher was the world's foremost chess player, he often demanded changes in the physical size of the boards and the pieces, the lighting and seating arrangements, and other such *conditions of play*.

Most westerners see a relatively clear distinction between the rules which define the game and the protocols which regulate the behaviors of the players in particular matches. Rules dealing with time limits and adjournments are not considered *rules of the game.*

Most board games have an implicit *touch-move* regulation which teaches children and other beginners never to *take a move back.* Yet, this rule does not prevent a teacher or game commentator from digressing off into some variations of play and then retracting back to the main line.

Go has had some protocol rules which are not well known in the west. It is considered rude to place the first stone in the opponent's lower right hand corner. So, there are old tales of a player placing his first stone in his opponent's right hand corner, and the referee then declaring the game forfeited because of this rude gesture! But fortunately, since everyone will correctly perceive that a westerner who makes such a play is hopelessly ignorant rather than rude, his move will be tolerated and he can remain blissfully ignorant of that unwritten rule of protocol.

Although a westerner might view a *pass* much like any other move, the protocols concerning it have often been quite special. Historically, some arrogant players have used a *pass* as an emphatic statement that they think the opponent's prior move is so bad that it requires no response. Correctly interpreting such a *premature pass* as a gesture of contempt, some referees have forbidden it.

In some traditions, the *pass* move has been partially exempt from the implicit *touch-move* regulation. Even if both players *pass*, one or the other may change his mind, and the game may *resume*, just as if the prior move or two were retracted. Some official rule statements include attempts to legislate the relevant protocols. Some of these *"rules"* appear to be courtesy guidelines, urging a victorious player to let his defeated opponent retract a pass that would yield an unduly lopsided score. But this also contributes to the inscrutability of the Japanese rules, whose passing and scoring methodologies also contain inherent logical intricacies which westerners find very difficult to decipher, even after discussions and explanations decouple the *resume play* protocols from the *confirmation phase* of post-passing play. This confirmation phase turns out not to be a playing phase at all, but rather an analysis phase, according to which various prospective lines of play might be tested if needed to resolve disputes about life, death, territory, and score. Fortunately, the positions which give rise to such disputes are sufficiently arcane that millions of games are played and resolved by Japanese players who have never read the fine print of the official rules.

A.7   References to official rules

The Go Player's Almanac [Boz92] includes five sets of official current rules: Japan, China, The Ing Foundation, New Zealand, and the North American Go Association. These rules appear in Chapter 13, preceded by an introduction by James Davies. John Power's history of modern Go, which includes some comments on how these official rules have evolved since 1868, appears in Chapter 5 of the same book. The now-obsolete 1986 rules of the Ing Foundation, which included the notion of prorating dame in seki, appear in [Ing86].

Further discussion of rules, and our efforts to mathematize them, appear in Appendix B.

A.8   Overview

In Appendix B, we expand this list of Section A.7 to include *ancient Go*, which was played in China circa 1100 AD, and several even more primitive rulesets that we define. These rulesets can be conveniently ordered as follows:

	Complex scoring rules
Japanese rules	*Early passing*
Modern non-Japanese rules	
Ancient rules	
Greedy ancient rules	
Mathematical rules	*Trivial scoring rules*
	Lengthy encores

All of these rulesets are surprisingly close to each other. Which ruleset is being used matters only in a rare position in a rare game, but a study of foundations necessarily emphasizes these differences.

The lower the position of a ruleset on this list, the longer a complex endgame might last. Japanese games never continue beyond moves which fill dames between independently live groups. We call any further play beyond that point an *encore*. All non-Japanese scoring systems naturally resolve any potential scoring disputes by letting play continue into the encore. Modern Chinese games may continue until dead stones are naturally captured and removed from the board. Ancient games continue until the losing side has filled in all of his territory except for two eyes per live group, and perhaps a few seki. And, in principle, if the losing player refuses to resign, Mathematical Go might continue on further still, until the only empty nodes on the board are single-node eyes of the winning player.

The higher a ruleset appears on this list, the greater the potential for a complicated terminal position, which may require more complex scoring rules. So it is no surprise to find that the rulesets which appear near the top of this list are much more complex than those near the bottom. Mathematical rules require no scoring at all; the game's termination and the loser are defined simultaneously when a player has no legal move. Values of earlier positions arise naturally from the mathematics. In non-pathological terminal positions, these values equal scores. Ancient rules allow passing, but ancient scores are defined simply by counting stones on the board. Modern Chinese scores are defined by counting stones and territory, but the territorial scores can be very simply defined because all stones which remain on the board in the terminal position can be presumed to be alive. Although there are differences among the modern non-Japanese scoring systems, they all belong on the same line of this chart, well below the modern Japanese rulesets and well above the ancient ones.

We have two goals in Appendix B.

1. We attempt to *axiomatize* the foundations of Go, as much as we can, by presenting simple postulates from which many of the basic notions and their properties can be logically derived. We do this by appending an artificial *greedy* rule to the ancient rules, yielding a very slightly modified game whose terminal positions are much more easily characterized.

2. We attempt to *mathematize* each of the modern rulesets, by defining for each a corresponding set of mathematical rules, in which passing is illegal. We also investigate the rare positions in which these mathematical rulesets differ with each other and the much rarer positions in which they differ from the corresponding popular games.

Appendix B

Foundations of the Rules of Go

B.1 Abstract

Our interest in foundations of rules is motivated by several considerations. First and foremost, we seek to obtain new results and a deeper understanding of real late-stage Go endgames by the application of the powerful methods of combinatorial game theory to Go. We would like our results to have as few limiting restrictions as possible. Combinatorial game theory is the study of sums of games in which the last move wins. Here is one of its most basic theorems:

> *If Black, moving second, can get the last move on a game A, and Black, moving second, can get the last move on a game B, then Black, moving second, can get the last move on the game A + B.*

The proof is immediate and constructive. With the aid of John Conway's axiomatic definitions, this theorem can be restated as follows:

> *If $A \geq 0$ and if $B \geq 0$, then $A + B \geq 0$*

The theory also includes many advanced theorems and powerful techniques for analyzing sums of games in which the last move wins. These techniques, which were introduced and described in ONAG [Con76], Winning Ways [BCG82], and [Ber88], include canonical forms, numbers, cooling, overheating, and warming operators. The fact that these techniques can now also be applied to Go was the primary factor in motivating us to write this book.

Late-stage Go endgames naturally decompose into sums of partial board positions, or *games*. But the full power of combinatorial game theory can be directly applied only to games in which the last move wins. And so we design rules of Go in which the outcome is decided not by score, but by who gets the last move.

It is not hard to engineer an exhaustive methodology which must always work in principle. One begins by considering the set of all relevant terminal Go positions, from which Black and White gurus would both choose to pass. One accepts whatever score their rules specify. This integer-valued score is then translated to an integer-valued combinatorial game in canonical form, and then used as the "foundations" on which to construct the values of nonterminal Go positions. Since tied outcomes are possible in Go but not in combinatorial game theory, one should evade them by including a half-integer valued komi, which is represented as a separate additive half-integer valued combinatorial game in canonical form. Unfortunately, the "foundations" required by this approach are large and cumbersome. They include a database of all relevant terminal positions. This is quite feasible for sufficiently simple partial board positions which have few terminal positions, but the quest for more generality leads to a quest for simpler scoring rules, which in turn leads to inspections of the foundations of all relevant rule systems.

Engineers and applied mathematicians strive to find mathematical models for a wide variety of real-world problems. Those who study the theory seek to formulate mathematical models of whatever sets of rules are used in major tournaments. The many different sets of Go rules used in different parts of the world today naturally divide into two principal classes: Japanese **J** and non-Japanese. One prominent example of the latter is North American, **NA**. In this appendix, we work towards the formulation of a mathematical ruleset **MJ**, which approximates **J**, and another mathematical ruleset **MNA** which closely approximates **NA**. In the course of these studies, we discover that the discrepancies in outcomes determined by **MJ** and **MNA** are significantly less common than one might naively expect. We introduce another mathematical ruleset **MU**, called *Universalist*, whose outcome often agrees with both **MJ** and **MNA**.

It is reasonable to presume that all of the modern dialects of rules arose from a common ancient historical ancestor. We begin this appendix by summarizing our favorite version of these ancient scoring rules, including their distinctive scoring rule: Ignore both territory and captives, and count *only* stones on the board. We reinforce this foundation for our subsequent studies by appending an additional *greedy* rule, which restricts passing in a way that is plausibly not disadvantageous. We are then able to give a complete characterization of the terminal positions of the greedy version of the ancient game. We then define Japanese and non-Japanese versions of the greedy ancient game, and give a complete characterization of the one-point scoring differences that can arise depending on who gets the last move. This possible one-point difference persists in modern versions of official Japanese and non-Japanese scoring systems, where it adds to a variety of other modern scoring differences. By studying it in the context of greedy ancient Go, we are able to isolate this feature from all of the modern scoring discrepancies which depend on details about what constitutes territory. In other words, we use ancient Go and greedy ancient Go to provide simplified contexts within which we can formulate and prove several facts about the similarities and differences between other classes of rule systems of more intrinsic interest, namely Japanese and non-Japanese, both popular and mathematical.

The following twelve terms denote important notions which are all well understood by experienced Go players:

liberty	atari	snapback
ko	string	group
seki	eye	false eye
big eye	dead stone	independently alive group

Many Go books define these terms only by examples. Sometimes these notions are defined in terms of each other in ways which a mathematically trained beginning Go player would consider circular. In this appendix, we give precise definitions to all of the above terms, at least in the simplified context of greedy ancient rules.

B.2 Ancient Go

The game has always been played by two players who alternately place black and white stones on the Cartesian lattice grid points of a board. Black moves first. (The first six moves of a Go game on a 7 × 7 game are shown in Figure B.1.) The players attempt to enclose territory and/or capture opposing stones while evading capture themselves. The primary rule, which defines how and when stones are captured and moved to a prisoner's pile off of the board, is identical for all rulesets considered in this book.

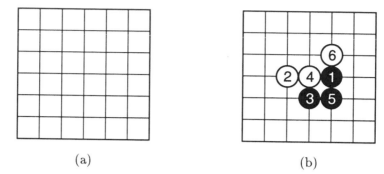

(a)   (b)

Figure B.1: *The first few moves*

B.2.1   The capturing rule

A *string* of, say, black stones is a set of stones of one color connected by the lines of the grid. So in Figure B.1(b), Black has a string of three stones, and White has strings of one and two stones. A *liberty* of a string is an empty node connected by a line to the string. If, after Black's move, a white string has no liberties, it is removed from the board. The captured stones are termed *prisoners* and are set aside for scoring. Naturally, White can also capture Black. If a Black move captures no stones, but leaves a black string with no liberties, the move is *suicidal.*

Examples demonstrating the capturing rules are shown in Figure B.2. White's play at *a* captures three black stones, which would then be removed from the board. White's play at *b* still captures the three stones, even though the move also leaves the white string with no liberties. The White play at *c* is suicide; it captures no black stones and leaves the white string with no liberties.

The White play at *d* captures a black stone marked ⬤, but Black can reply by playing at ⬤ capturing 3 White stones. This sequence of 2 consecutive moves is called a *snapback.* The first move of a snapback captures a single stone. The second move of the snapback captures several stones of the opposite color by moving onto the node just vacated by the captured stone.

Lastly, suppose Black plays at *e*, capturing the stone labeled Ⓐ in the corner. If White could immediately recapture, the position would repeat, and Black and White could recapture one another forever. This is called a *ko*, and all nonmathematical rulesets prohibit this recapture at least in

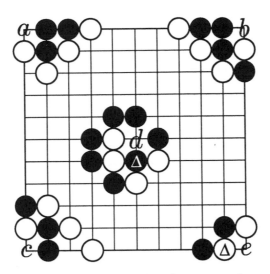

Figure B.2: *Examples of capture rule*

the case where the repeated position occurs every 2 moves. However, White could choose to recapture after both players have played elsewhere, since the full board position would no longer repeat. This is by far the most common case, where the repetition is caused by a single stone capture. Other repeated positions are extremely rare.

Sometimes a ko might be very important to the outcome of a game. For instance, the life and death of a large group might depend on who gets to capture. When this is the case, and a player is prohibited from recapturing the ko, a *kofight* ensues. Both players try to make forcing moves elsewhere just for the sake of changing the board position so they can take back the ko. Such forcing moves elsewhere are termed *kothreats*.

B.2.2 Ancient rules

We now summarize all of the rules of ancient Go:

0. **Initial Conditions:** *We start with a single empty board, and no captives.*

1. **The Capturing Rule:** *(Details are given in the prior section)*

2. **Suicides:** *Any suicidal move is illegal.*

3. **Ko Playing:** *Any move which repeats any previous board position is illegal.*

4. **Passing:** *Instead of playing a stone on the board, a player may take a stone of his color from the pot and give it to the opponent as a captive. This move is called a pass.*

5. **Termination:** *Play continues until either player chooses to resign* **or** *both players agree on the final score* **or** *there are two consecutive passes, with either player passing last.*

6. **Scoring Philosophy:** *My* **score** *= Stones on board of my color.*

7. **Outcome:** *The player with the greater score wins. Ties are possible.*

B.2.3   Historical note

The version of the rules we describe as *ancient* was played in China circa 1100 A.D., possibly with some minor modification of rules 2 and 3. The game originated in China at least 2000 years earlier, but there is considerable doubt about what ruleset was originally used.

Substantially different versions of Go have been played in Tibet and Sikkim, and it is not known when that version split off from the rulesets discussed in this book. In Tibet each side begins with a few special immortal stones (called *bulls*) on the board as part of the specified initial conditions. But a more profound difference is a modification of the capturing rule to prohibit snapbacks. This might be viewed as a generalization of the ko rule. Instead of prohibiting the immediate recapture of a single stone only when needed to avoid loopy positions, in Tibet they banned it in all circumstances. One profound implication of that rule change is that it invalidates Section B.5. We do not consider the Tibetan dialects of Go any further in this book. The interested reader is referred to [Fai90a], [Fai92], [Sho93].

Shotwell's article also includes a substantial discussion of the relationships between game rules and the philosophies of the cultures wherein they flourished.

B.2.4   Greedy ancient Go

We now wish to explore the terminal positions, in which a competently played ancient game might end. To this end, we seek to formulate the conditions under which good ancient players would both elect to pass. Here is one plausible such condition, which we state as an amendment that we will arbitrarily annex to the other ancient rules.

4+. Greedy Passing Restriction: *A player is not permitted to pass unless both of the following statements are true:*

1. *He has no legal move which immediately captures some opposing stone(s).*

2. *If he makes any move on the board, the opponent can immediately respond by capturing the string which includes the just-played stone.*

Even if Rule **4+** is not included, then a player might nevertheless voluntarily adopt a strategy which also satisfies Rule **4+**. Such a strategy involves reading[1] one move ahead, and refusing to pass unless the projected ancient score (= net number of stones on the board) is optimized. Since this strategy maximizes the score over a very short-term horizon, we call it *greedy*. The rules of *greedy ancient Go* coincide with the rules of ancient Go, except that the passing rule is amended to Rule **4+**.

B.2.5 Groups of stones and topological configurations

Let us now study the Terminal Greedy Ancient Positions, which we abbreviate TGAPs.

We may classify all TGAPs in terms of the topological relations between their empty nodes and their strings. To do this, we construct a *full graph* whose nodes correspond to the TGAP's empty nodes, and whose branches correspond to the TGAP's strings. Each branch in this graph is colored black or white according to the string it represents. Two branches of the same color are said to be *parallel* if they connect the same pair of nodes. All of the stones in a set of strings which correspond to a set of parallel branches are said to be a *group* of stones. We also form a *condensed topological graph* by combining each set of parallel branches in the full graph into a single branch in the condensed graph. There is then a one-to-one correspondence between nodes in the condensed graph and empty nodes in the TGAP, and a one-to-one correspondence between branches in the condensed graph and groups of stones in the TGAP.

---

[1]Computer scientists and others familiar with chess use terms like *lookahead*, which are synonymous with the term *reading* as used in Go jargon.

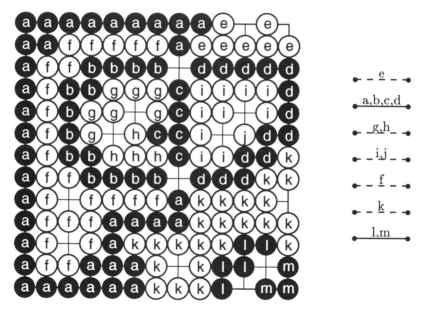

Figure B.3: *As many as four strings can share life in a TGAP*

Figure B.3 shows that it is possible to have as many as four parallel strings in a single group. This is indeed a highly contrived position. In a real professional game, the occurrence of a group consisting of even two strings separated by a small live opposing group is considered noteworthy. (See "Living with a false eye" Nakayama Noriyuki 6-dan [Nak91].)

In addition to a simple black group in the middle and a moderately common configuration in the lower left, the TGAP in Figure B.4 includes a highly contrived example of a configuration whose condensed topological graph is a linear string of 9 branches and 10 nodes. Each of the upper corners shows how a pair of parallel strings can contain only one stone each.

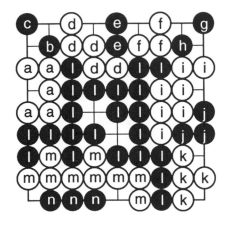

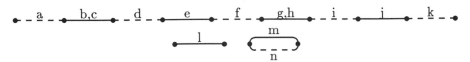

Figure B.4: *Bichromatic linear seki with 9 groups and 11 strings*

B.2.6 Decomposition of a TGAP into a sum of configurations

We may partition the condensed graph into the union of components, each of whose nodes is connected to each other by branches within that component. Such a connected component is called a *configuration*. Each configuration must either be linear or cyclic. Here is the complete list of all possible configurations.

- Monochromatic:

- Bichromatic Linear:

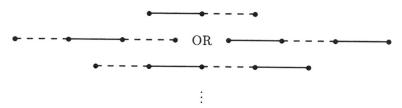

$$\vdots$$

- Bichromatic Cyclic:

B.2.7 Proof of completeness of the list of configurations

We are now able to prove that every TGAP is a sum of these basic configurations. The converse is not true; there are various exceptions of increasing complexity which we will investigate in Section B.5.

The fact that every configuration of every TGAP is one of the above forms follows from the following elementary assertions:

- Every string in a TGAP has exactly two liberties.

Proof: If a black string had only one liberty, the greedy white player would capture it rather than pass. If it had three or more liberties, the greedy black player would play on one of those liberties rather than pass. ∎

- In the full graph, any monochromatic pair of branches which share one node necessarily share both nodes.

Proof: If two black branches violated this condition, then instead of passing, the greedy black player would play on the unique node which the two branches shared. That move would combine the two strings into one, whose liberties would include the node at the other end of each former string. Since those nodes are different, the new string would have at least two liberties. ∎

B.2.8 Definitions of *seki* and *independent life*

Our greedy passing rule has quickly yielded a necessary condition which tightly constrains all of the possible terminal positions. Each such terminal position is a disjoint sum of configurations. Each monochromatic configuration is called *independent life*; each bichromatic configuration is called a *seki*.

Notice that a TGAP partitions every node of the entire Go board into such configurations.

B.3 Local versions of the ancient rules

As evidenced in Section A.3, all modern rule systems emphasize territory, although they quibble among themselves over precisely how territory is defined. This is entirely a modern issue, since the ancient rules we have stated make no mention of territory.

Modern Chinese and Japanese rule systems also differ with each other over the philosophical issue of whether to count stones on the board or prisoners. We find it convenient to study this issue independently of the issue of territorial definition. The way we prefer to do this is to project the modern differences back onto the ancient rules.

We begin this study by considering some rules that were adopted by the North American Go Association in 1992. We state these **NA** rules as minor amendments to our ancient rules. We include the corresponding **C** rules for comparison. All other rules remain as stated in Section A.3.

B.3.1 Philosophical revision of the ancient rules

$$
\begin{aligned}
\mathbf{C} &= \text{modern China} \\
\mathbf{NA} &= \text{North American}
\end{aligned}
$$

5. Termination: *Play continues until either player chooses to resign* **or**

> **5b. NA:** *Black passes and then White's next move is also a pass.*
>
> **5c. C:** *There are two consecutive passes, with either player passing last.*

6. Scoring Philosophy 6a. C: *My* **score**= *Stones on board of my color.*

> **6b. NA:** *My* **score**= *Captive stones of opponent's color.*

We conventionally define the net score from Black's perspective:

$$\text{Net score} = \text{Black's score} - \text{White's score}$$

B.3.2   American score = Chinese score

Although the American ancient rules and Chinese ancient rules differ in philosophy, we shall now show that they have no difference in substance. To this end, let

$$
\begin{aligned}
\mathbf{C} \;&=\; \text{ancient Chinese score} \\
&=\; \text{Black stones on board} - \text{White stones on board} \\
\mathbf{NA} \;&=\; \text{Americanized ancient score} \\
&=\; \text{White stones captured by Black} \\
&\quad\; - \text{Black stones captured by White}
\end{aligned}
$$

But we notice that in either set of rules, each player must take one stone from his pot at each turn. At any point in the course of the game, each stone that has left the pot is either on the board or in the pile of captives. Since Black made the first move, after each White move each player has taken the same number of stones from the pot, and we have the identity

$$\mathbf{C} = \mathbf{NA}$$

So, as long as we ensure that White makes the last move, there can be **no difference** in score.

Rule **5b** ensures that score-counting occurs after a White move, even if one or both players made earlier intermittent pass(es) before the final game-ending pair of passes.

B.3.3 Komi

Go experts consider the privilege of starting to be worth about 5 to 8 points. Hence, in modern tournaments it is very common to declare that all final scores will include an adjustment called a *komi*. The size of this komi may vary from tournament to tournament. Although some tournament directors define the komi to be zero, the most commonly used value is −5.5 points. This means that Black must pay a 5.5 point penalty in return for the privilege of making the first move.

Tournament authorities have traditionally abhorred ties. Often they eliminate them by specifying a non-integer value of the komi. We henceforth assume that the komi is such a value. This is mathematically very convenient, because classical mathematical game theory does not readily allow tied outcomes.

Many problems are stated in the form, "White to move and win." Although the komi is nominally zero, the statement implies that White should not consider a tie to be a satisfactory outcome. Therefore, we should take the value of the komi, K, as $+\frac{1}{2}$.

We now consider a Japanese version of the ancient rules, and three mathematical versions. The North American = Chinese version is also included, for comparison.

B.3.4 Five variations on the ancient rules

$$
\begin{aligned}
\mathbf{J} &= \text{Japan} \\
\mathbf{NA} &= \text{North American} \\
\mathbf{MU} &= \text{Universalist (A class of mathematical rulesets)} \\
\mathbf{MJ} &= \text{Mathematized Japanese} \\
\mathbf{MNA} &= \text{Mathematized North American = Mathematized Chinese} \\
\mathbf{M} &= \text{all of the three prior (mathematical) dialects}
\end{aligned}
$$

Each of these five rule *ancient* rule systems also has a *greedy ancient* variation, obtained by including Rule **4+** of Section B.2.4.

0. **Initial Condition** *with K point komi. K* **must** *be of the form $E + f$, where E is an even integer, and f is a fraction with $|f| = \frac{1}{2}$.*

   0a. **J, NA, MJ:** *The appropriate player is given K stones as initial captives.*

   0b. **MU:** *The appropriate player is given E stones as initial captives.*

0c. MNA: *The komi is rounded up to* $K + \frac{1}{2}$, *an integer, and the appropriate player is given that many initial captives.*

4. Passing

4a. M: *Instead of playing a stone on the board, a player may return one captive stone of the opponent's color to the pot. If the player holds a half-stone captive of the opponent's color, he may return it to the pot instead of a full-stone. If his opponent holds a half-stone captive, then the player may instead replace his opponent's half-stone captive with a full-stone captive.*

4b. NA, J: *Instead of playing a stone on the board, a player may take a stone of his color from the pot and give it to the opponent as a captive.*

5. Termination *Play continues until either player chooses to resign* **or**

5a. M: *One player has no legal move, in which case the game ends and he loses.*

5b. NA: *Black passes and then White's next move is also a pass.*

5c. J: *There are two consecutive passes, with either player passing last.*

6. Scoring 6b. NA, J: *My* **score**= *My Komi* + *Captive stones of opponents' color*

B.3.5 Mathematized ancient rules

The three systems of mathematical rules included in Section B.3.4 embody a substantially different philosophy from the others. They require a player to return a prisoner at each pass. The game ends when the loser has no more prisoners to return, and is unwilling or unable to play on the board.

If passing is illegal, then one might expect that the losing player would feel compelled to make a costly sacrifice move from a TGAP. Of course he is permitted to do so if he wishes, in which case the game would continue until his only potential moves are illegal suicides, because his opponent so totally dominates the board [Tak90]. However, mathematically, from a TGAP in which Black has a nonnegative net score, a costly sacrifice move by White is no better than no move at all. Suppose, for example, that the score is 2, and White's only move changes the score to 15, but that Black has a move which changes it to 1. Then combinatorial game theory assures us that

$$2 = \{1|15\} = \{1| \ \}$$

which is a formal way of saying that White's only canonical move is to resign.

So, in our Universalist version of greedy ancient rules, we can make the canonical assumption that play on the board ends when the position reaches a TGAP. The value of the position is then given by a number U, which is equal to the net difference between the numbers of total prisoners of the two colors, including both those that were captured on the board and those that were included in the initial komi. The winner of a Universalist greedy ancient game is Black if $U > 0$, White if $U < 0$, and last player to move on the board if $U = 0$.

B.3.6 Parity

In this next section, which independently mimics Section 3.6, we show that there is a relationship between which player moved last, the parity of U, and the parity of the number of odd sekis in the TGAP.

If we examine the board and count the captives, then we can determine whose turn it is to move next. Since Black played first, it is Black's turn if an even number of moves have been made and it is White's turn if an odd number of moves have been made. Each move takes one stone from a pot, and each such stone is either on the board or in the captive pile. Hence,

Number of turns = number of stones on board + number of captive stones

The number of stones on the board may be calculated as the total number of nodes on the board minus the number of empty nodes on the board. In a TGAP, there are very few empty nodes on the board. In live groups and in cyclic sekis, all such empty nodes come in pairs. Only some types of linear sekis have an odd number of empty nodes; we call such sekis *odd* sekis. We let S denote the total number of odd sekis. Then in a TGAP on an odd×odd board,

$$\text{Number of stones on board} = 1 + S \pmod 2$$

We notice that U, which is the difference in the numbers of white and black captives, has the same parity as the sum of the numbers of captives. Therefore,

$$\text{Number of turns} = 1 + U + S \pmod 2$$

Since we have represented the komi, K, as $K = E + f$, and E is an even integer, E, has no affect on these parity arguments.

B.3.7   One-point adjustments among C, J, and MU

For greedy ancient games, we now consider the conditions under which the outcome determined by our ruleset **MU** agrees with the outcomes determined by **J** or **C** = **NA**. The relevant parameters are,

- S, the number of odd sekis in the TGAP

- who moved last

- U, the net number of prisoners including E, the even part of the komi.

- f, which is $+\frac{1}{2}$ or $-\frac{1}{2}$ depending on the parity of the komi

The tabulation of all of these cases is presented in Figure B.5.

According to Section B.3.6, the parity of U depends only on S, the number of odd sekis in the TGAP, and on who made the last move. The four relevant cases are tabulated in row 1 of Figure B.5. Row 2 states the ranges of U which are necessary and sufficient for Black to win, in accordance with the conclusion of Section B.3.6. It is more convenient to restate these conditions in a form which compares U to an integer of opposite parity; this is done in row 3.

If we instead use the ruleset **J**, the greedy ancient outcome is independent of S and of who got the last move; it depends only on the net difference in the numbers of prisoners, including the komi $K = E + f$. Since the value of U already includes E, but not f, the relevant necessary and sufficient condition for Black to win is now $U + f > 0$. Equality is impossible because f is not an integer. This condition, rephrased as $U > -f$, appears in the fourth row of Figure B.5. In the fifth row, we list the conditions for Black to win according to the **MJ** rules. These differ from the fourth row only in that Black might win by getting the last move with a tied score. But since U is an integer and f is not, there is no actual difference between the conditions on rows 4 and 5. **MJ** = **J**.

If White got the last move before the TGAP, then rulesets **NA** = **C** award Black the same net score as ruleset **J**. However, if Black got the last move on the board, then rulesets **NA** = **C** award Black one point more than ruleset **J**. In cases when Black gets the last move, she wins in ruleset **C** if and only if $U + f + 1 > 0$. This condition appears in the appropriate

	The number of odd sekis, S, is even		The number of odd sekis, S, is odd	
Who moved last $\longrightarrow$	White	Black	White	Black
1. Parity of U	odd	even	even	odd
2. Black wins **MU** iff	$U > 0$	$U \geq 0$	$U > 0$	$U \geq 0$
3. or equivalently,	$U > 0$	$U > -1$	$U > 1$	$U > 0$
4. Black wins **J** iff	$U > -f$	$U > -f$	$U > -f$	$U > -f$
5. Black wins **MJ** iff	$U > -f$	$U \geq -f$	$U > -f$	$U \geq -f$
6. Black wins **NA** iff	$U > -f$	$U > -f-1$	$U > -f$	$U > -f-1$
7. Black wins **MNA** iff	$U > -f-\frac{1}{2}$	$U \geq -f-\frac{1}{2}$	$U > -f-\frac{1}{2}$	$U \geq -f-\frac{1}{2}$

Conditions for discrepancy in outcomes				
8. If $f = +\frac{1}{2}$	$0 \geq U > -\frac{1}{2}$	$-\frac{1}{2} \geq U > -\frac{3}{2}$	$1 \geq U > -\frac{1}{2}$	$0 \geq U > -\frac{3}{2}$
i.e.,	impossible	impossible	$U = 0$	$U = -1$
9. If $f = -\frac{1}{2}$	$\frac{1}{2} \geq U > 0$	$\frac{1}{2} \geq U > -\frac{1}{2}$	$1 \geq U > \frac{1}{2}$	$\frac{1}{2} \geq U > -\frac{1}{2}$
i.e.,	impossible	$U = 0$	impossible	impossible

In case of discrepancy				
Value of f	impossible	$-\frac{1}{2}$	$+\frac{1}{2}$	$+\frac{1}{2}$
rulesets favoring Black	none	**MU, NA**	**J, NA**	**NA**
rulesets favoring White	none	**J**	**MU**	**MU, J**

Figure B.5: *Summary of one-point adjustments to Mathematized endings* columns of row 6 of Figure B.5. On row 7, corresponding to **MNA**, the value of the komi is increased by $+\frac{1}{2}$, and the conditions for Black to win are enlarged to include cases when the score is tied and Black gets the last move on the board. However, since f is a half-integer, the conditions on row 7 are soon seen to be identical to the conditions on row 6. For greedy ancient games, our mathematized versions of **J** and **NA** are perfect models, **MJ** = **J** and **MNA** = **NA**, and so we need further investigate only three of our five greedy ancient rulesets. We select **MU**, **J**, and **NA**. In order to obtain any discrepancy in the outcomes determined by these three rulesets, the value of U must lie between the maximum and the minimum of the three lower bounds shown on rows 3, 4, and 6. The range of potential discrepancy is shown on row 8. From this we deduce the value of U, if any, which corresponds to any discrepancy. The exercise is done separately for $f = -\frac{1}{2}$ and $f = +\frac{1}{2}$. The conclusions appear in the last few rows of Figure B.5.

Although we have developed Figure B.5 in the simplified context of greedy ancient endgames, it will later become apparent that the conclusions persist when all of the relevant rulesets are modernized. In the modern versions, the occasional one-point discrepancies tabulated in Figure B.5 will appear *in addition* to various other (rare) modern scoring discrepancies, such as those illustrated in Section A.1.

When $f = +\frac{1}{2}$ and if S is even, then, according to Figure B.5, rule systems **MU**, **J**, and **C** = **NA** all yield the same ancient greedy outcome, independent of who got the last move. This is a very common case. In many endgames, $S = 0$. A substantial majority of modern professional games end without any sekis in the terminal position. And the most common type of seki is a standoff between two opposing no-eyed groups which share two liberties. This is an even seki, which is not counted in S. An odd seki in a TGAP must be a bichromatic linear configuration, such as a standoff between two one-eyed groups which share one liberty. The value $f = +\frac{1}{2}$ is also very common. The value of komi which has been most widely used in Japanese tournaments in the past decade has been $K = -5\frac{1}{2} = E + f$, where $E = -6$ and $f = +\frac{1}{2}$. All of our endgame problems in Appendix C of this book are stated as, "White to move and win." This implies that a tie should be considered an unsatisfactory outcome for White. In other words, we should consider ties as favorable to Black, and this is most readily done by setting the komi as $K = f = +\frac{1}{2}$. All of these endgame problems also have $S = 0$. Thus, these endgame problems are universal, in the sense that it usually doesn't matter whether we score according to Japanese rules, or

Chinese rules, or North American rules, or Universalist rules.[2] In every case, White wins if, and only if, he can get the last one-point move. Of course, the Mathematical Universalist rules provide a rigorous logical foundation for the solutions of these problems, as given in Appendix D.

B.3.8 The mathematized half-integer komi

A half-integer included in the Japanese komi is represented mathematically as a special initial half-captive. The **MJ** rule for dealing with such a half-captive is taken from the standard rules of mathematical game theory. Specifically, if n is a positive integer, then

$$n = \{n-1 \mid \}$$

This is taken to mean that if Belle Black has n captives, then she may, at her turn, return one to the pot, whereas Wright White has no option to do anything other than to play on the board. Similarly,

$$n + \frac{1}{2} = \{n \mid n+1\}$$

This is taken to mean that when a half-stone captive is present, then either side has the option, in lieu of a move on the board, to alter the net number of captives by a half-stone in favor of the opponent. After one such move, the special half-stone captive is out of play; only normal full-stone captives remain.

B.4 Modeling by mathematical rules

B.4.1 Foreseeing the terminal configurations

We have seen that every terminal position of a greedy ancient game decomposes uniquely into disjoint configurations. Each configuration is either *independent life* or *seki*.

Competent players can often foresee the terminal configurations long before they reach the TGAP. Indeed, the point in the play when the boundaries of the terminal configurations become stabilized is precisely the point at which modern Japanese games terminate. But then the ancient game

---

[2]To ensure universality, we must also evade other discrepancies in modern scoring systems. The only problem in Appendix C in which this becomes a significant issue is Figure C.19.

continues quite a while longer. In the final stages of the ancient game, the players alternately fill in territory they have already surrounded. From the modern perspective, the ancient game appears much like a childish school performance. The most interesting part is really over, but the performers seem determined to continue playing an encore, which may go on until they reach a TGAP.

There are often a very large number of ways to play an encore flawlessly. That is surely one reason that experienced players consider most encores to be so dull.

B.4.2   Dead stones in ancient territory

At the beginning of the ancient encore, a live black group may enclose a few harmless white stones. During the encore White may attempt to turn these stones into a small live group, or to create a seki, but a competent Black player can thwart all such attempts. Empirically, if it were not clear that Black can succeed in capturing all such white stones during the encore, then the white stones would not be *harmless*; the topology and boundaries of the TGAP would not yet be clear. Modern Chinese players would not yet have passed; the ancient encore would not yet have begun.

These harmless stones are said to be *dead*. This usage of the word is unique to Go. In Go, *dead* means "destined to be captured," during the encore if not sooner.

Combinatorial game theory assigns mathematical values to every game position. Unless someone poses an imminent threat, an ancient encore position's mathematical value will be an integer, equal to the net number of moves that Belle can make before she must pass. Mathematically, the value of the incentive of an integer is -1. That means that either player who moves changes the value by one point against himself.

It is convenient to view the value of a live black region plus the value of the captives that it will yield as a single entity. The value of this entity is evidently given by the formula

$$V = 2D + E + C - 2$$

where

$$E \quad = \quad \text{number of empty nodes in the region surrounded by Black}$$
$$D \quad = \quad \text{number of dead White stones in the region enclosed by Black}$$
$$C \quad = \quad \text{net number of White stones captured by Black from this region}$$

Indeed, from the perspective of combinatorial game theory, it is sufficient to verify that at least one of the following is true:

$$V \;=\; \{V-1 \| V \mid -\text{BIG}\}$$ White has an immediate kothreat, but Black has an adequate response.

$$V \;=\; \{V-1 \mid V+1\}$$ Most common case.

$$V \;=\; \{V-1 \mid \;\}$$ All remaining empty nodes are one-point *eyes*.

In general, any noncapturing move by Black reduces E and V by one. A move which captures X stones yields the new value of

$$V' = 2D' + E' + C' - 2 = V - 1$$

because

$$
\begin{aligned}
D' &= D - X \\
E' &= E + X - 1 \\
C' &= C + X
\end{aligned}
$$

So any Black move on V yields the new value, $V' = V - 1$.

White's move is to play another stone into the region, decrementing E while incrementing D. The net result is to increment V. Occasionally White may have a move which captures Y black stones. As long as this move has no affect on the overall life of the region, then it yields only

$$V' = 2D' + E' + C' - 2$$

where

$$
\begin{aligned}
E' &= E + 1 + Y \\
C' &= C' - Y \\
D' &= D + 1
\end{aligned}
$$

whence $V' = V + 1$.

Sometimes White's move may pose a grave threat, which would lead to a value of $V'' = -\text{BIG}$ if Belle failed to respond immediately and Wright were permitted to make the next play in this region. But, assuming that Belle has an adequate response, then the net affect of White's threat and Black's response is to increment D by one while decreasing E by two, yielding the same value of V as before that consecutive pair of moves occurred.

Eventually Belle will capture all of the dead white stones and reduce all remaining territory to isolated eyes. White will then have no more legal plays in this region, because his attempt to play into an isolated Black eye is a one-stone suicide, which is illegal in all versions of the rules. Black can continue playing, until the net number of Black moves in this region and the captives resulting from it is V. At that point the only remaining unfilled nodes in the region will be the last two eyes which Black needs to retain life.

This verifies the asserted formula for the ancient value of a black group surrounding a territory containing D dead white stones, E empty nodes, and C net associated captive white stones. The formula is

$$V = 2D + E + C - 2$$

An interesting property of the formula is that it may also be written as

$$V = (E + D) + (C + D) - 2$$

Evidently, at any time during an ancient encore,

> The value of a position is unchanged if dead stones are removed from the board and treated as captives.

We have derived this fact as a consequence of the ancient rules. Since it is often presented as a rule rather than as a result, it creates a befuddling amount of confusion among beginners.

The formula for the ancient value, V, also reveals another amusing fact:

> The value of a position is unchanged if live black stones within black territory are removed from the board and treated as captives, or, conversely, if black prisoners are used to fill black territory.

The removal of each live black stone increments E but decrements the net value of C. Of course, one can also arbitrarily remove live white stones from within white territory, or use white prisoners to fill white territory.

The results we have just derived can be applied to *any* of the ancient rule systems we have considered: Americanized, Japanized, or Mathematized. This result also shows that, under any set of rules, the score of a competently played ancient encore is independent of the details of how it is played out. The crucial assumptions are that the terminal configurations and their boundaries are precisely known at the beginning of the ancient encore, and that either player can play the ancient encore in such a way that protects his claims. Subject to these conditions, any additional details of play can have no effect on the final score.

B.4.3 Modernized territory in living regions

Many combinatorial games have natural values equal to the net number of moves which Belle can make from that position. For a Go position consisting of an empty region enclosed by Black, this natural value is its territory $- 2$. That agrees with ancient rules rather than with modernized rules. To create a mathematized rule system whose results correspond to modernized rules rather than to ancient ones, it is necessary to find a way to ensure that the net number of moves Black can play on the board is equal to the full amount of her territory. The obvious way to do this is to allow Belle to fill her last two eyes without adverse consequences. This can be accomplished by introducing the notion of *earned immortality*, as an amendment to the capturing rule.

B.4.4 Earned immortality

1a. All modern mathematical rulesets: *Whenever either side has a parallel set of strings that each has exactly two liberties, such that every node adjacent to each of these two liberties is occupied by some member of this set of strings, then those two liberties are said to be "a pair of eyes." The group consisting of all parallel strings which adjoin this pair of eyes is "promoted to immortality." This promotion can be recorded by drawing an extra line onto the board, which connects this group to a special "node at infinity." Since the node at infinity can never be filled, the group can never thereafter be captured, even if both of its eyes are filled.*

B.4.5 Ko-playing, a la Conway

Both the modern Japanese and modern western versions of the rule for playing loopy positions appear in Section A.3, rule **3** (page 116).

Either rule can be easily mathematized, but both lead to relatively cumbersome mathematics. For some (limited) purposes, another rule option proves very attractive mathematically:

3': *A move which repeats the position that occurred two moves ago, or any other prior board position, is* **legal**.

We call 2-move loops *Conway loops* in honor of John Conway, who pioneered the study of *loopy games* in Chapter 11 of Winning Ways [BCG82]. The obvious disadvantage of such a definition is that it leads to many hung outcomes. But the advantage is that it leads to a huge increase in tractability. It preserves complete independence of summands. It allows all considerations involving kothreats to be ignored. It is far more general than specifying that the board contain no kos or prospective future kos. The Conway Ko-playing rule is implicitly identical to a condition which has long been used by traditional Go players in partial board endgame problems: If possible, win without relying on Ko. If you can't do that, then win relying on Ko. And if you can't do that, then resign. One can logically restate this as saying that game outcomes are ordered:

$$WIN > HANG > LOSE$$

B.4.6 Ko playing, a la Japanese *confirmation phase*

In the rarely used *confirmation phase* of 1989 Japanese rules, the ban on recapturing in a ko remains in effect even after both players play elsewhere on the board. The ban can be lifted only by passing. If there are several kos on the board, then each time a player passes, he must specify which stone he will be allowed to retake.

In current practice, confirmation phase rules are used only for analysis, not for actual play. After the game has stopped, there are rare occasions when the score may depend on intricate subtleties of the mutually intertwined and peculiarly Japanese notions of life, death, and territory. Under such circumstances, several further lines of continued play under confirmation phase rules may be relevant to resolving such questions. We'll say more about the modern Japanese rules in Section B.5.7.

These rules provide another interesting way to decouple the values of positions involving kos from distant kothreats. For that reason, they merit further mathematical study.

B.4.7   Chinese komi

One easy way to model a komi in (non-mathematical) Chinese scoring is to start not from an empty board, but from the sum of two boards, one of which is the standard empty 19 × 19 board, and the other of which is shown in Figure B.6.

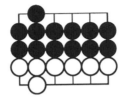

Figure B.6: *The 6 point Chinese komi*

B.4.8   Mathematical rules which approximate the Japanese ko-scoring rule

Suppose that one player, say Black, surrounds a point in a ko that is active when White first passes. Then Japanese scoring assigns no territorial point to this node. It turns out, on careful inspection, that all other rule systems we consider agree with the Chinese on this issue, and assign one point to Black. And so there arises the problem of finding a way to amend the mathematical rules to bring them into closer agreement with the Japanese on this issue.

We address this issue by modifying the initial conditions. In addition to the initially empty board, we also include the board shown in Figure B.7. This board ensures a plentiful supply of dame. These dame inhibit the possibility that, in almost any competently played game, either player might choose to pass (which incurs a one-point penalty) during an active kofight.[3]

Since the boards of Figures B.6 and B.7 each has an even number of nodes, neither appendage has any affect on the parity arguments of Sections 3.6 and B.3.6.

B.4.9   Encore formally defined

The ruleset **MJ**, which represents our current closest mathematical approximation to Japanese rules, includes the half-stone komi described in Sec-

---

[3]For some very rare contrary examples, see Section B.5.6.

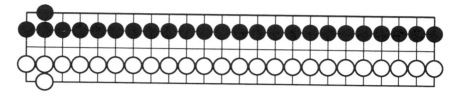

Figure B.7: *A 2n × 5 board providing 2n dame*

tion B.3.5. We can imagine any endgame problem or position played according to our rules **MJ**, by two omniscient gurus,[4] each of whom plays flawlessly. Then we define the start of the encore as the move which resolves the half-stone captive.

There is a crude correspondence between our notion of *encore* and the *confirmation phase* which appears in the 1989 version of Japanese rules. All dame between independently live groups will be played before the encore begins.

In non-Japanese scoring systems, there is no compelling need to define the start of the encore formally, for it can make no difference in outcome. To gain a better understanding of why this is so, we next investigate some of the positions that typically occur in games shortly before the onset of the encore, when some dame may still be present on the board.

B.5   Traditional basic shapes

B.5.1   False eyes and sockets

A false eye, as shown in Figure B.8, is a shape which beginners often mistake for an eye. In each situation, we assume all stones are connected to neighboring strings with liberties.

As in Section 4.7 we view a false eye as a *socket*. This is a point at which a connection between two strings is likely to be made. Figure B.8(d) shows one of many cases in which the strategic shape is not yet resolved. If White

---

[4]We also use the similar term *competent* to describe flawless play. The distinction between a *competent player* and a *guru* only reflects the authors' estimate of the difficulties to be encountered. We describe a flawless sequence of play as competent if, in our opinion, most experienced Go players ranked 10-kyu or higher could be expected to find it in reasonable time. The reader who wonders why we invoke gurus for so seemingly mundane a purpose as defining the onset of the encore should study the rest of this appendix with care.

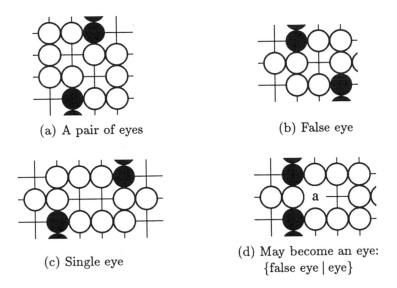

(a) A pair of eyes   (b) False eye

(c) Single eye   (d) May become an eye:
{false eye | eye}

Figure B.8: *Some basic eye (and false eye) shapes*

plays at *a*, he creates an eye, but if Black plays a sacrificial stone there, its capture creates a false eye, or socket.

B.5.2 Sequences of sockets

Figure B.9 shows a sequence of four distinct black sockets appearing along the top edge of a 9 × 9 board. This sequence of sockets terminates in black strings *B* and *C*, which continue into unshown lower regions of the board, as does the white string *A*.

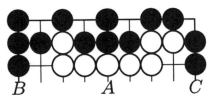

Figure B.9: *Four sockets (or false eyes) along the top*

If *B* and *C* are connected, then all black stones shown become part of a single live group, and the 4 sockets become 4 points of modern territory.

The next table shows some of the possibilities which can occur if B and C are disjoint. To reduce the number of cases, we assume that B is at least as strong as C:

A	B	C	Status of sockets and adjacent strings
2 eyes	≥ 1 eye	≥ 1 eye	4 points of live territory
2 eyes	2 eyes	0 eyes but liberties	Must be filled, 0 points
	seki		Depends on rules

In the case when these groups form a seki, the modern score of the territorial points in the sockets depends on the rules. An example of such a situation appears in Figure A.2 of Section A.2.

Japanese rules give no points for the sockets in the seki. All other rules, both modern and ancient, score them as one point of territory each. Since each such node provides a place where Black can move but White cannot, its natural value is given by a fundamental equation of mathematical game theory, namely:

$$\{0|\} = +1$$

Japanese rules evade this equation by outlawing the encore.

B.5.3 Defective shapes in TGAP-imposters

We now consider some matters that form an important and essential ingredient of basic Go strategy. These issues occur frequently, and the conclusions reached in this section are universal, for all rulesets considered in this book.

Not every configuration which might appear to be a TGAP actually is one. Even when our greedy passing rule permits greedy ancient players to pass, there is no requirement for them to do so. If a competent greedy ancient player can find a sacrifice move which will improve his eventual score, we can assume that he will play it. A configuration which succumbs to such a sacrifice attack is not a genuine TGAP; it is *defective*.

Let us now examine some conditions under which a competent greedy ancient player can improve his score by making a sacrifice move. Without loss of generality, we assume that the sacrificing player is White.

The sacrificing move will connect all strings in its group into a single string, whose shape we call the *sacrificed shape*. This string then necessarily shares a single liberty with one opposing group. We assume that Black accepts the sacrifice by playing onto this liberty and capturing the sacrificed string. This capturing move by Black may unite several Black strings into a single capturing group. White then has the opportunity to invade the empty region which has the sacrificed shape. This region is entirely surrounded by Black. Under some conditions this situation seems very favorable to Black. We state maxims that advise White not to get into such a situation:[5]

Maxims on when not to sacrifice:

- Never sacrifice any sufficiently large (say ≥ 10 stones) group.

- Don't sacrifice a group if the Black strings adjacent to it belong to more than one Black group.

These restrictive maxims permit White to make a small sacrifice against a single targeted black group. After such a sacrifice, the targeted black group will necessarily enclose a targeted empty region of the sacrificed shape, but it will have no external liberties.

Whether White can now successfully attack the targeted black group depends entirely on the sacrificed shape. Figures B.10 and B.11 include some shapes which White can successfully attack. These diagrams show the shape of the sacrificed white string at the time it is captured.

Figure B.10: Single eye *shapes. If a single black string captures either of the above shapes, Black immediately has 1 eye.*

Black's capture transforms the sacrificed white string into an empty region which has the same shape as the white string just captured. White should immediately play onto a key central point of that empty region. If Black ignores this invasion, then on subsequent moves, White expands his invading string. This expansion continues until it again threatens to capture the targeted black group. Black can evade the imminent capture only by capturing the invading white string, or by making some earlier play in the region. In either case, the size of the region which the targeted black string encloses shrinks in size, and remains one of the shapes shown in the Diagrams. White's invasion resumes or continues until Black plays again. After at most a few such Black moves, the targeted region reduces to a single

---

[5]We view *maxims* as *semi-rules* and place them in a type font intermediate between that used of formal rules and that used for text.

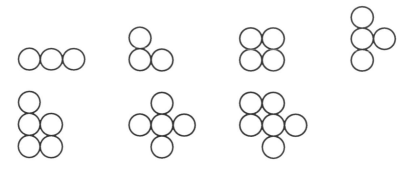

Figure B.11: Big eye *shapes. If a single black string captures any of these shapes, Black can be assured of only one eye.*

node, from which White captures the entire targeted black string. White's final capturing move is often a snapback.

Other shapes lead to different results. Figure B.12 shows the empty "four in a line" regions which White *cannot* successfully invade. Each line of four empty nodes has two central nodes, and if White invades either, Black can successfully defend by playing the other. Black has an even easier time defending an empty line of length ≥ 5.

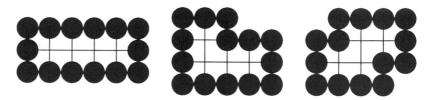

Figure B.12: *Four in a line is a living eye space*

This completes our preliminary study of defective shapes. Here is a summary of the conclusions, as they relate to nominal TGAPs.

1. A connected monochromatic configuration is defective only if their two empty nodes are adjacent. In all other cases the two empty nodes are said to form *a pair of eyes* and the configuration is *independently alive*.

2. A two-group configuration is not defective if the potential sacrificing group is adjacent to more than one opposing group. This is true even if the two empty nodes are adjacent. (Figure B.13)

3. A two-group configuration is defective if the group to be sacrificed is adjacent only to the targeted group, and if the sacrificing group can form a big eye.

4. A two-group configuration is not defective if the sacrificing group forms ≥ 4 empty nodes in a line in the interior of the board.

5. A connected configuration with more than two groups is never defective.

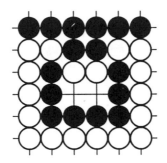

Figure B.13: *A 2-group seki with adjacent empty nodes*

B.5.4 Semi-terminal ancient positions

Even before the position has reached a TGAP, the possibility of a defective shape may leave one or two empty nodes upon which only one side can play. Such a position is shown in Figure B.14, which we have taken from [KOM86]. The position is destined to be a seki. White can play at nodes *a* and *b* whenever he wishes. Black cannot play on either node, for to do so would create a defective shape allowing White eventually to capture all black stones shown.

So this seki position is terminal for Black, but not for White. We call it *semi-terminal*. It behaves just like the false eyes in seki, which occurred at nodes *L*, *M*, and *N* of Figure A.3. In non-Japanese scoring, these nodes become points for the player who can play there. But in Japanese scoring, no score is awarded for such points.

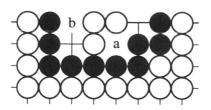

Figure B.14: *A semi-terminal seki position*

B.5.5   Other terminal ancient positions

There are a few uncommon examples of positions in which a competent ancient player would choose to pass even though the greedy rule states otherwise. Figure B.15 shows two such positions. Evidently, each of these is a *terminal ancient position* even though it is not a *terminal ancient greedy position*.

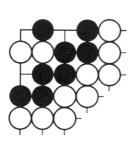 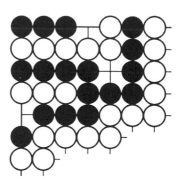

Figure B.15: *Terminal ancient positions which are not termed terminal greedy ancient positions*

Since these positions are terminal, in any mathematical ruleset their values are zero. In popular Go jargon, both of these positions are called *sekis*. The position on the right is called a *hane-seki*.

B.5.6   Bent 4 in a corner

The list of 7 big eye shapes shown in Figure B.11 is complete, up to rotations and reflections. No other such shapes can occur in the interior of the board, or along an edge. If a single black group surrounds any empty region which is either slightly or moderately larger, then the targeted black group cannot be captured, even if White plays first.

However, Figure B.16 shows two other *big-eye-like* shapes which can occur in the top left corner. The smaller shape is *Bent 4 in a corner*; the bigger 3 × 2 rectangle in the corner is another shape which can lead into it.

Figure B.17 shows that after Black captures the shape of Figure B.16, she can survive only by winning a kofight in the corner. And thus the famous *bent 4 in a corner* shape gives rise to some fascinating situations.

Figure B.16: *Above left (in the top left corner of a Go board) is a bent 4 in a corner shape. Above right might lead to bent 4.*

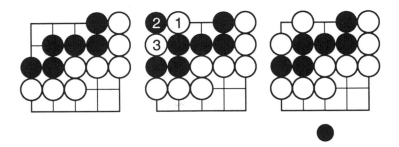

Figure B.17: *Bent 4 in the corner. After 3 plays it's a kofight, and Black must make the first kothreat.*

In Figure B.18, the encore has just begun. Even though Black has many kothreats and White has none, Black's group in the upper left of the board is dead, under all rulesets. White patiently puts off capturing it as long as possible. He first fills in all of his own territory except the last two eyes, at a cost of one point per move. But Black must also pay one point per move, either by likewise filling in her territory, or by conceding an extra captive each time she passes. Only after White has reduced his own territory down to the minimum pair of eyes does he play at *a* which attacks Black with the bent 4-in-a-corner shape. Black captures at *b*, but then White plays as in Figure B.17. At this point Black has no kothreats, and cannot prevent White from capturing on his next turn.

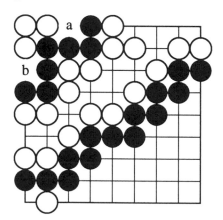

Figure B.18: *White wins by 1 (easily counted by Chinese scoring along the main diagonal)*

This argument generalizes to the famous maxim:

Bent 4 maxim:

- Bent 4 in a Corner Kills, independent of the rest of the board.

As shown in Figure B.18, there is a real case to be made in favor of this maxim. Current kothreats are irrelevant. The attacker can eliminate all of them before he begins the kofight.

This maxim has appeared explicitly in many versions of Japanese rules. The Japanese 1989 *confirmation* rules have been designed to retain the truth of this maxim without stating it explicitly.

But the position in Figure B.19(a) demonstrates how this maxim can be incompatible with the results of a competently played encore.

If Black's upper left string is considered dead, then White wins by 3, and this would be the verdict of a Japanese referee. However, when played out according to rules of any other country, White's quandary becomes clear. The seki on the right side of the board offers either player an irremovable kothreat. After White has reduced his own territory to only two eyes, we then see the sequence of Figure B.19(b–d). Then White must choose between

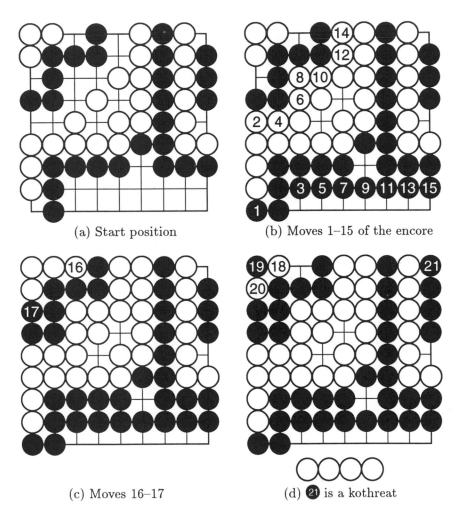

(a) Start position

(b) Moves 1–15 of the encore

(c) Moves 16–17

(d) ㉑ is a kothreat

Figure B.19: *In (a), White appears 3 points ahead (because of bent 4 in the corner), but he loses the encore by 7. Diagrams (b–c) show how when Black has an irremovable kothreat, the bent 4 in the corner may not be dead.*

killing Black's string on the left, or saving his own on the right. Either way, White loses by 7 points.

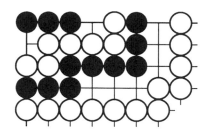

Figure B.20: *A more complex variant of bent 4 in the corner*

Figure B.20 shows yet another interesting variation. White threatens to fill 6 liberties and capture the 9 black stones, ensuring life for himself and death for the 3 black stones in the corner. The only way that Black can prevent this is to capture the 7 white stones first. To do that, Black must start and win the corner kofight while her 9-stone group has at least 2 liberties. In order to win that kofight, Black may have to ignore a small White kothreat, and so she should postpone starting the corner kofight until she has eradicated as many White kothreats as she can. This is likely to be her best strategy even if some of the moves needed to eradicate such White kothreats have a mathematical incentive of -1. These are moves which are not normally played until the encore. And while Black is making such moves, White will be filling the liberties, which from a normal (ko-free) point of view might plausibly appear to be "dame."

If there is a *Bent 4 in a corner* shape on the board, the encore and the stage of play just before the encore can become quite exciting!

This shape is likely to become a major strategic factor in any game in which it occurs. The reader may enjoy studying two professional Japanese games in which shapes of this sort appeared [OK91] [Sak81, p. 173]. In [OK91], the shape appears early in the game, in a context similar to Figure B.20. As in Figure B.20, Black's group had no eyes, but it had sufficiently many liberties that there was enough time to remove **all** of White's kothreats, at a cost of several points, before capturing the white group in the corner to survive and win the game by a comfortable margin. In [Sak81], the kofight appeared only very late in a very close game. One player, who seemed destined to win the game by $+\frac{1}{2}$ point, could do nothing better than to fill a one-point ko and a dame, while the apparently losing player started a corner kofight. This drastically altered the contest, changing the apparent half-point victory into a big defeat.

B.5.7 Positions which stress the fine print of Japanese rules

The logical difficulties and complexities of the official Japanese rules have attracted ample criticism from Japanese authors, including several 9-dans [KOM86]. A thorough and excellent exposition entitled "Defects of the Japanese Rules" appears in Part II of a booklet written by Ikeda Toshio, a very strong amateur Go player who was also an inventive and colorful mainframe computer pioneer at Toshiba. The English translation of this book [Ike92] was published posthumously.

In Section A.8, the Japanese rulesets appeared at the top of the list. A Japanese game might end a few moves sooner than the same game would if played according to any other ruleset, in which play might continue into an encore. But, such encores are rare in all modern rulesets, and when they occur, they are usually very short. Japanese rules abolish all such encores at the cost of a large increase in complexity of rules. In Japanese rules, after all dame between independently live groups are filled, any further move typically costs one point of territory. Therefore, Japanese players are incentivized to pass while there are still dead stones on the board. In some rare cases, the players may express different views about whether stones are alive or dead. Some such disagreements cannot be resolved by continued normal play, because further moves would have adverse effects on the mover's score. In this section, we discuss some of the rare positions whose scores depend on subtleties of the Japanese rules.

There are positions from which competent Japanese players will both pass even though one or more kos are still active. According to the official rationale for scoring such positions (described in Section B.4.6) the seven black stones in the upper left of Figures B.18 or B.19 are all dead. But, according to intricate official Japanese analysis which we shall not repeat here, the eight black stones in the upper left of Figure B.21 are alive, and the lone white stone is dead. In all of these cases, the dead stones may be removed from the board and taken as prisoners.

There are other arcane terminal positions which arise so rarely that discrepancies and ambiguities in the official rules can (and do) persist for decades.

But, although most of the world has learned Go from Japan, all other

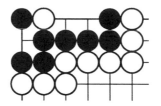

Figure B.21: *The approach ko*

countries except Korea have officially rejected official Japanese rules in favor of something else which they consider more logical, less ambiguous, and/or much simpler for beginners to learn [Lai90]. Ikeda[Ike92] and other Japanese authors have also proposed revised rules which address these same concerns.

Traditionally, Japanese scoring disagreements were resolved on a case by case basis as they arose. By 1949, the list of positions requiring special official explanations included over 40 examples. In 1989, the Nihon Ki'in adapted a new set of official Japanese rules, which attempted to eliminate the special precedents and reduce the rules to a single legal page. These rules appear on pages 230–231 of *The Go Player's Almanac* [Boz92], but they are followed by another 11 pages of official commentary and examples of how these rules should be applied. It is rumored that the original, more detailed, draft of these rules was significantly longer. Its adoption could not be reconciled with the myth that, "Go has simple rules." This concern was addressed not by changing any substance, but by compressing the commentary and explanation to the official 1989 version, which some experienced scholars have been able to interpret only after extensive oral discussions with members of the Nihon Ki'in rules committee.

Non-Japanese scoring systems define rules and scoring in terms that beginners can understand and apply. All of them are compatible with the relatively simple definitions of *life* and *death* that we have stated earlier: A stone is *dead* if it is destined to be captured in a competently played encore. All non-Japanese rulesets are also compatible with the notion that all dead stones can be removed from the board without affecting the score. However, the official 1989 Japanese rules entail substantial complications of these concepts.

Figure B.22 shows one of the positions included in the official explanations. The official score of this position is 0. The official explanation is that, "The 7 black stones and the 1 white stone are both dead." Yet neither officially dead group can be removed from the board without affecting the score. Since the two empty nodes are surrounded only by officially dead stones, there is officially no territory. So we christen this position an *anti-seki*.

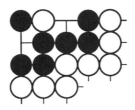

Figure B.22: *An anti-seki with score 0. Under 1989 official Japanese rules, seven black stones and one white stone are officially dead, but cannot be removed.*

Part of the official explanation as to why these *dead* stones cannot be removed is that, "White has no territory." This reliance on *white territory* to distinguish the position of Figure B.22 from those of Figures B.23 entwines the official notion of *territory* with *life* and *death*. These official definitions are tied together in a knot whose logic is not easily untangled. In Japanese officialese, even if a stone can be captured, it may be *alive* if its capture, "would enable a new stone to be played that the opponent could not capture." [Boz92, p. 230, Article 7] The enabling mechanism is defined only by examples. A more formal definition would require examination not only of a hypothetical extension of play which exhibits the new stone, but also a contrary line of play which shows that any such new stone **could** have been captured **if** the *live* group in question had not been captured.

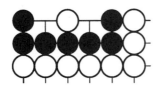 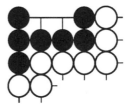

Figure B.23: *Seven dead black stones can be removed, in any ruleset*

As a practical matter, these confusing rules often have less impact than a logician might fear. For example, the main point of the official example we show as Figure B.22 is that, with competent play, this position is definitely **not** terminal in any ruleset discussed in this book. Black should capture the white stone before she passes, and she then has 3 points. So the precise value of the score of Figure B.22 as officially defined may be of only secondary importance, because it will matter only in an imperfectly played game. Any game between two gurus must necessarily end in a position which is terminal in whatever ruleset they are using.

The positions of Figure B.15 are also among those which receive official explanation. These are terminal under any non-greedy ruleset. For that reason, under any of our mathematical rulesets, their value is 0, which corresponds with the official score. Not surprisingly, Figure B.14 also has an official Japanese score of 0, although its Chinese score is -2. The value is -2 in all mathematical rulesets we have yet contrived, including **MJ**, which is our best mathematical approximation to official Japanese rules. Perhaps further amendments of the mathematical rules can be devised which will further reduce the small number of terminal positions in which mathematical values fail to match the official Japanese scores.

The official Japanese rules make no attempt to score some positions; they simply state that under some circumstances, "both players lose." The known examples are all nonterminal positions, from which competent players would not choose to pass. For that reason, one can argue that the fine points of Japanese rules have little impact in these cases. However, in a few fascinating and significant cases, the relevant position is terminal under some current or historical version of official rules.

Figure B.24 shows a simple position in the top left corner of the board. Each side takes his own view of the six nodes near the corner. White claims that one black stone is dead, and Black claims that four white stones are dead.

We first consider the simplest case, in which neither side has any kothreats. Then, in each of the two sequences shown in Figure B.24, every move prior to the last move is sufficiently threatening that it is reversible in the formal mathematical sense. So, canonical players will continue to play until the sequence is completed. After White's opening move, the full sequence shown leads to one dame on the board, which has the mathematical value $*$. Black also has 3 net captives, so the total value is $+3*$. After Black's opening move, the sequence leads to no points on the board, plus 2 net captives,

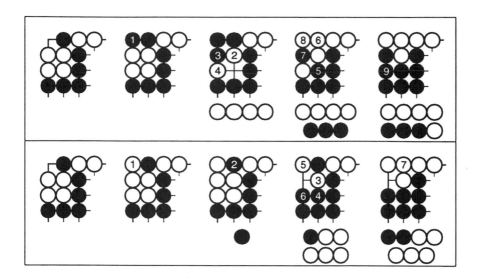

Figure B.24: *An unusual situation. In the first line of play, Black moves first. In the second, White moves first. Captives are shown below the boards.*

giving a total value of +2. So, mathematically we write the formal equation,

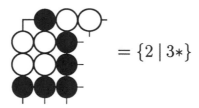 $= \{2 \mid 3*\}$

We now see why the position has been contentious. It is a rare example of a Go position which is apparently not yet resolved, but in which **either** player does **better** if his **opponent** plays first. Although we have assumed that neither side has any kothreats, a closer inspection reveals that the potential kofight is a further possible disadvantage to whichever player starts. If played according to our rules **MJ**, this position would not be played prior to the encore. Even in the encore, neither player should play it until he has filled in enough territory elsewhere to ensure that he has eliminated as many opposing kothreats as he can. From combinatorial game theory, we know that the canonical form of

$$\{2 \mid 3*\} = +3$$

Therefore, the net contribution of this position will be three points for Black. In a competently played close encore, Black will be the one who makes the first move in this position.

The starting position in Figure B.24 appears in the 1989 Japanese rules as the first example of confirmation of life and death. The explanation states that this position is seki. Its official score is 0. One consequence of these official Japanese rules is that the position is nonterminal. Black will play it before she passes, and attain a score of 2.

However, earlier Japanese authorities had once adjudicated this position and decided that it was terminal, with a score of 3. The position became a classic, with the title, *Three Points Without Capturing*. This title still appears as the section heading in the official 1989 examples, even though the 1989 rules now contradict the traditional title.

All non-Japanese rules and all mathematical rules agree with the verdict of the earlier Japanese adjudicators.

Appendix C

Problems

This Appendix contains problems which can be used for practice. Appendix D provides the values for each region from which you can deduce dominant winning lines using the information summarized in Figure E.9. Possible winning first moves are also indicated in the solution diagrams. Keep in mind, however, that simply knowing the first move is only a very small part of truly understanding the solutions to most of the problems. You can try out your solutions against Chen's software [Che] discussed in Section 1.4.

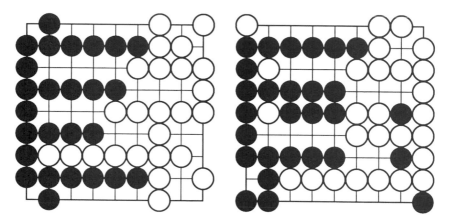

Figure C.1: *Numbers warmup*   Figure C.2: *Ups and stars warmup*

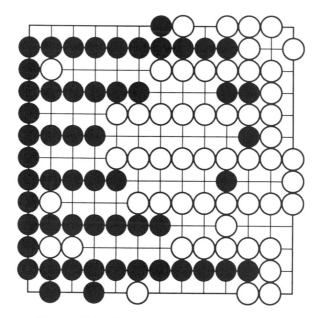

Figure C.3: *Combination of $0^n|x$ warmup*

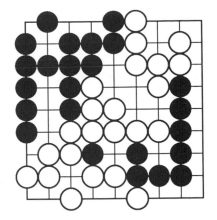

Figure C.4: #'s, ↑'s and *'s

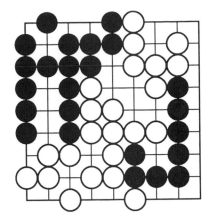

Figure C.5: #'s, ↑'s and *'s

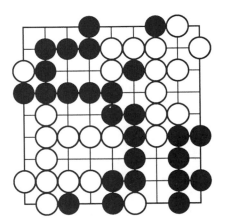

Figure C.6: #'s, ↑'s and *'s

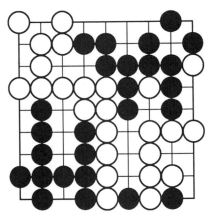

Figure C.7: #'s, ↑'s and *'s

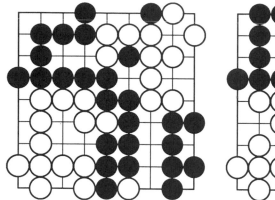

Figure C.8: #'s, ↑'s and *'s

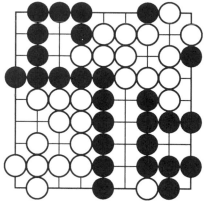

Figure C.9: #'s, ↑'s and *'s

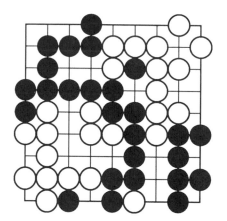

Figure C.10: #'s, ↑'s and *'s

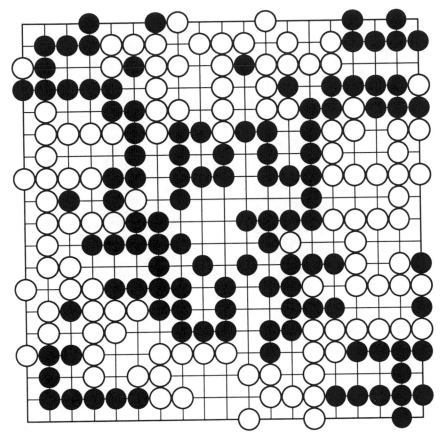

Figure C.11: #'s, ↑'s and *'s

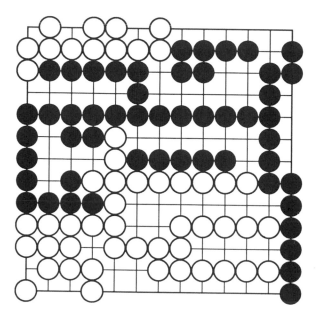

Figure C.12: $0^n | x$'s

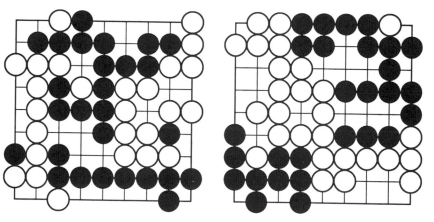

Figure C.13: $0^n | x$'s   Figure C.14: $0^n | x$'s

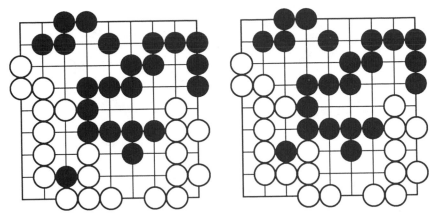

Figure C.15: *More ↑'s and *'s*   Figure C.16: *More ↑'s and *'s*

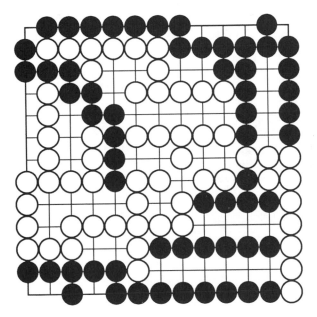

Figure C.17: *Multiple corridors*

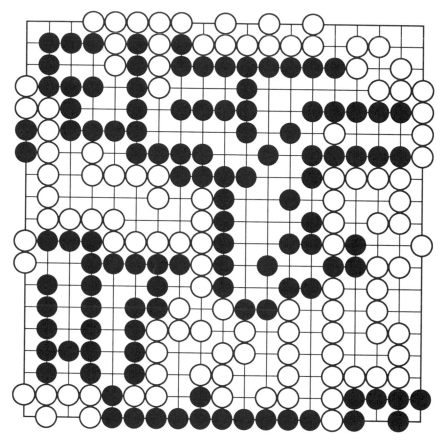

Figure C.18: *Multiple corridors*

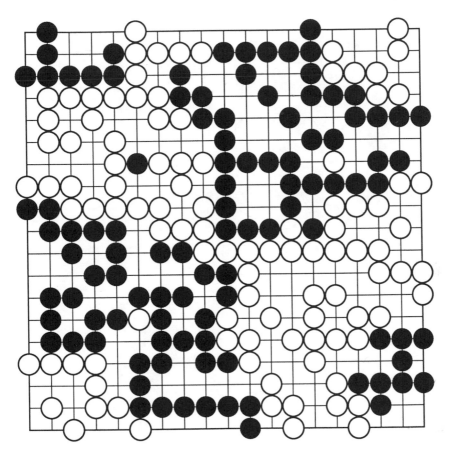

Figure C.19: *Lots of $0^n|x$'s*

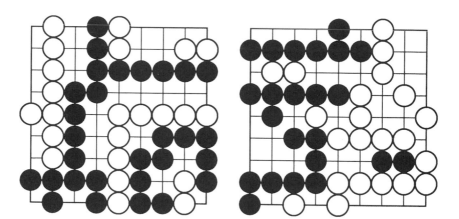

Figure C.20: *Capturing race? Five white captives are off the board.*

Figure C.21: *Hot miai*

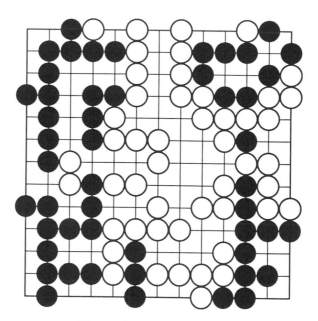

Figure C.22: *Two switches*

Appendix D

Solutions to Problems

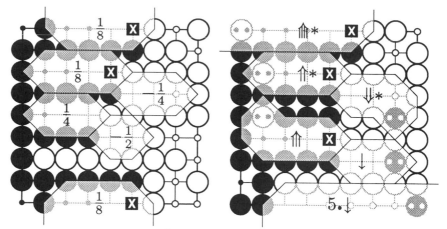

Figure D.1: *Total* $-\frac{1}{8}$ Figure D.2: *Total* $\uparrow*$

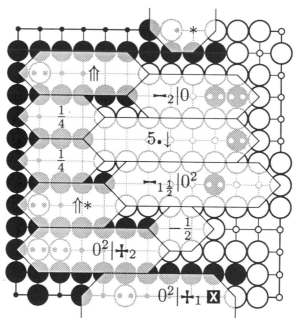

Figure D.3: *Atomic weight* 1

$\boxed{\mathbf{X}}$: Winning move(s)

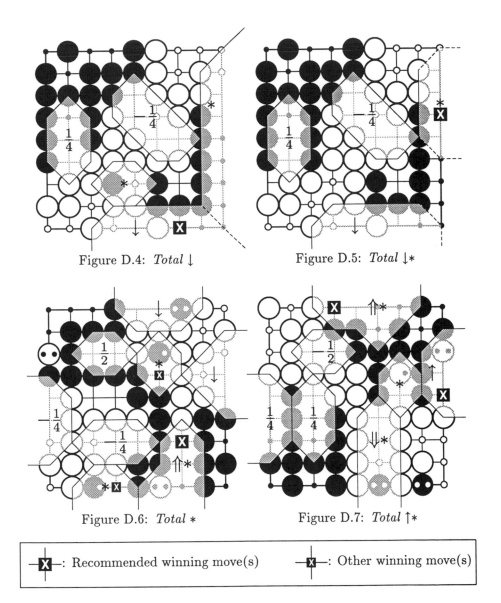

Figure D.4: *Total ↓*

Figure D.5: *Total ↓∗*

Figure D.6: *Total ∗*

Figure D.7: *Total ↑∗*

X─: Recommended winning move(s)   **x**─: Other winning move(s)

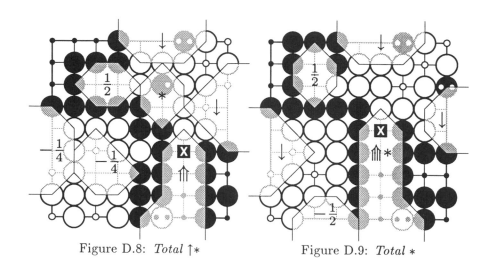

Figure D.8: *Total* ↑∗   Figure D.9: *Total* ∗

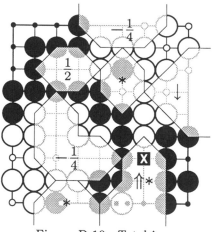

Figure D.10: *Total* ↑∗

—☒—: Winning move(s)

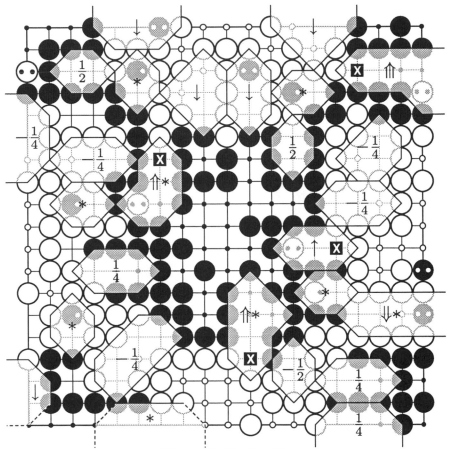

Figure D.11: *Total* ↑∗

| ─**X**─ : Winning move(s) |

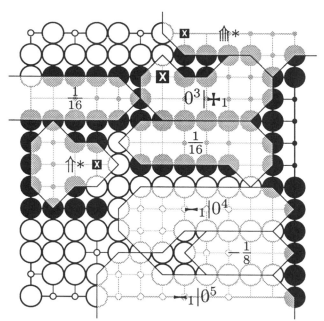

Figure D.12: *Atomic weight* 0

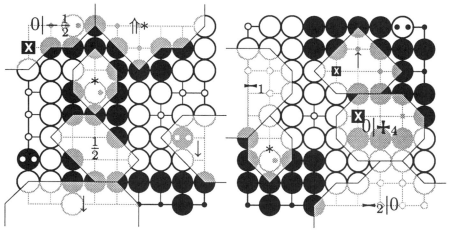

Figure D.13: *Total* $\frac{1}{2}|0$ Figure D.14: *Atomic weight* 1

X—: Recommended winning move(s)   **X**—: Other winning move(s)

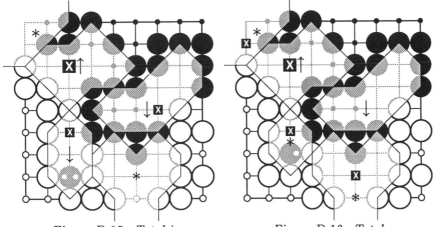

Figure D.15: *Total* ↓   Figure D.16: *Total* ∗

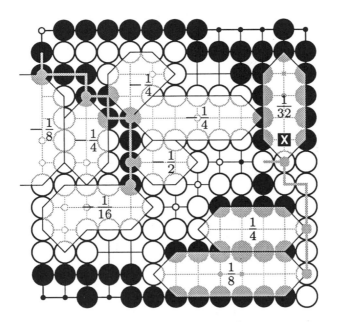

Figure D.17: *Total* $-\frac{1}{32}$

▗X▖–: Recommended winning move(s)	–▗x▖–: Other winning move(s)	

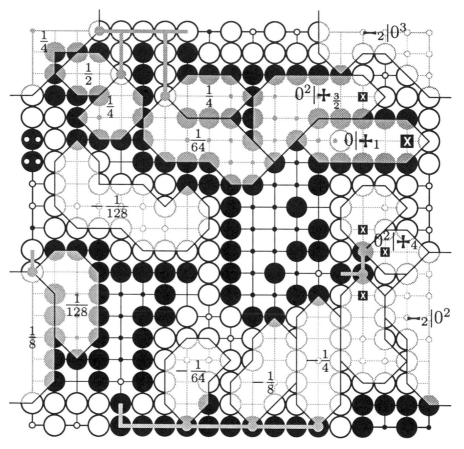

Figure D.18: *Atomic weight* 0

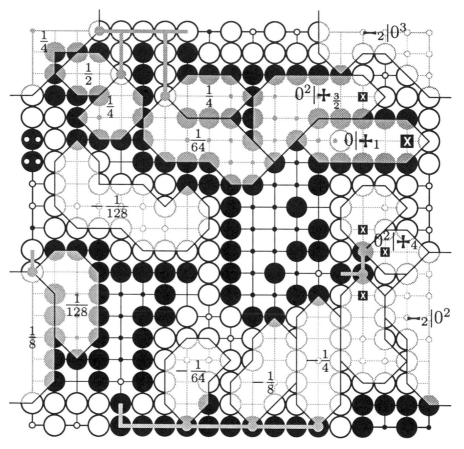 : Recommended winning move(s)   : Other winning move(s)

187

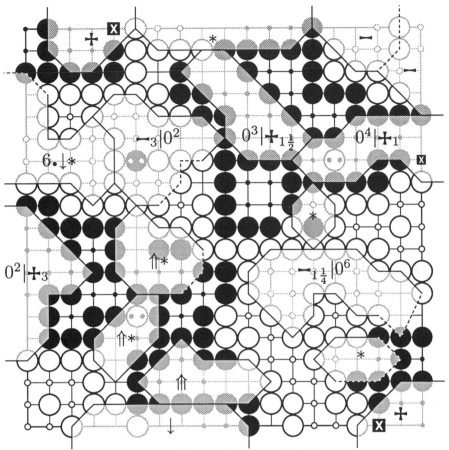

Figure D.19: *Atomic weight* 1

| ![X] : Recommended winning move(s) | ![x] : Other winning move(s) |

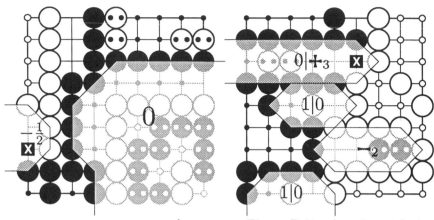

Figure D.20: *Total* $-\frac{1}{2}$
(There are 5 white captives)

Figure D.21: *Atomic weight* 1

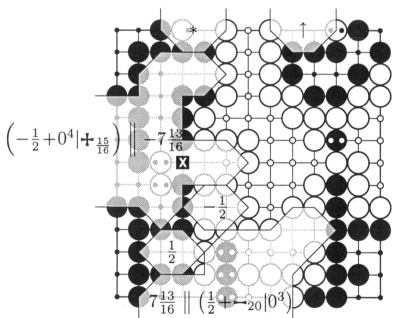

Figure D.22: *Only one switch works*

$-\boxed{\textbf{X}}-$: Winning move(s)

Appendix E

Summary of Games

Contents

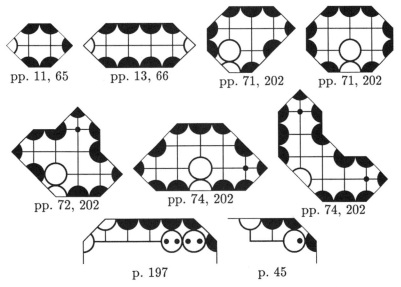

pp. 11, 65   pp. 13, 66   pp. 71, 202   pp. 71, 202

pp. 72, 202   pp. 74, 202   pp. 74, 202

p. 197   p. 45

Figure E.1: *Some examples of positions worth $\frac{1}{2}$*

189

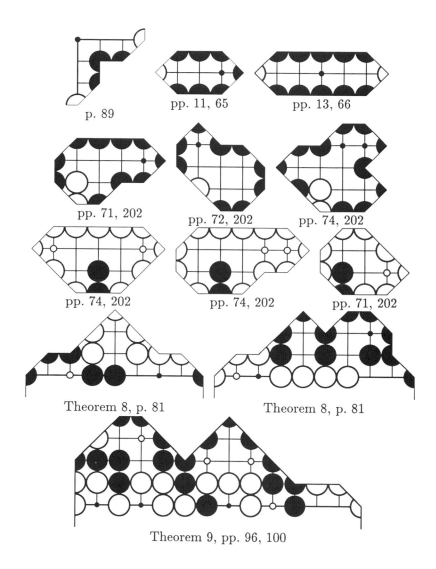

p. 89

pp. 11, 65

pp. 13, 66

pp. 71, 202

pp. 72, 202

pp. 74, 202

pp. 74, 202

pp. 74, 202

pp. 71, 202

Theorem 8, p. 81

Theorem 8, p. 81

Theorem 9, pp. 96, 100

Figure E.2: *Some examples of positions worth* $\frac{1}{4}$

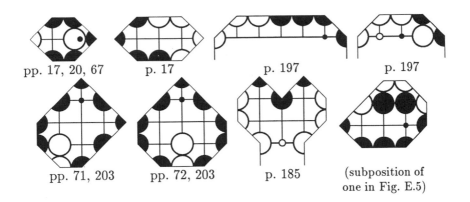

Figure E.3: *Some examples of positions worth* ∗

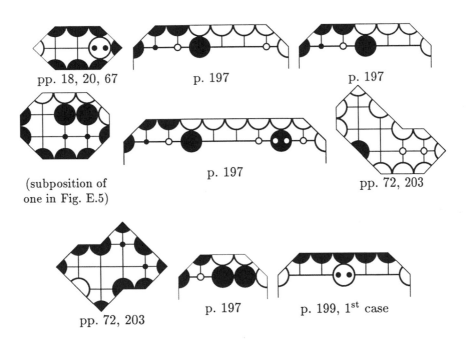

Figure E.4: *Some examples of positions worth* ↑

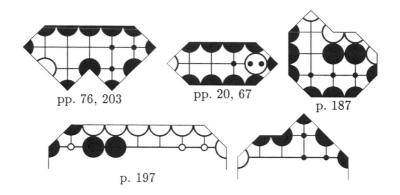

Figure E.5: *Some examples of positions worth* ⇑*

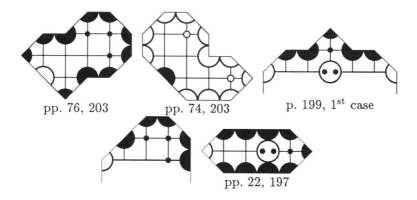

Figure E.6: *Some examples of positions worth* +₁

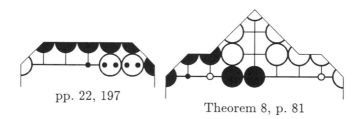

Figure E.7: *Some examples of positions worth* 0|+₂

GAME DEFINITION	PRONOUNCED	SECTION
$0 \overset{\text{def}}{=} \{\mid\}$	zero	3.5.3
$n \overset{\text{def}}{=} \{n-1\mid\}, \ n > 0$	positive integers	
$-n \overset{\text{def}}{=} \{\mid -n+1\}, \ n > 0$	negative integers	
$\frac{m}{2^k} \overset{\text{def}}{=} \{\frac{m-1}{2^k} \mid \frac{m+1}{2^k}\}, \ m$ odd	numbers	2.1, 3.5.3
$* \overset{\text{def}}{=} \{0\mid 0\}$	star	2.4, 3.5.3
$\uparrow \overset{\text{def}}{=} \{0\mid *\}$	up	2.4, 3.5.3
$\downarrow \overset{\text{def}}{=} -\uparrow = \{*\mid 0\}$	down	
$\uparrow * \overset{\text{def}}{=} \uparrow + *$ $\quad = \{0, *\mid 0\}$	up-star	2.4, 3.5.3
$n \boldsymbol{\cdot} \uparrow = \uparrow \boldsymbol{\cdot} n \overset{\text{def}}{=} \overbrace{\uparrow + \uparrow + \cdots + \uparrow}^{n}, \ n \geq 2$ $\quad = \{0 \mid (n-1)\boldsymbol{\cdot}\uparrow + *\}$ $\quad = \{0 \mid (n-1)\boldsymbol{\cdot}\uparrow *\}$	double-up, triple-up, ...	2.4, 3.5.3
$n \boldsymbol{\cdot} \uparrow * \overset{\text{def}}{=} n \boldsymbol{\cdot} \uparrow + *, \ n \geq 2$ $\quad = \{0 \mid (n-1)\boldsymbol{\cdot}\uparrow\}, \ n \geq 2$	double-up-star, ...	2.4, 3.5.3
$\Uparrow \overset{\text{def}}{=} 2 \boldsymbol{\cdot} \uparrow$	double-up	2.4, 3.5.3
$\boldsymbol{+}_G \overset{\text{def}}{=} \{0 \| 0 \mid -G\}, \ G > r > 0$ for $\boldsymbol{-}_G \overset{\text{def}}{=} \{G \mid 0 \| 0\} \quad$ some number r	tiny G miny G	2.5, 3.5.3
G_1	G chilled	2.2, 3.6
$\int G$	G warmed	3.6
$0^n \mid G \overset{\text{def}}{=} \underbrace{0 \|\|\| 0 \|\| 0 \| \cdots}_{n \text{ 0's}} \mid G$ $-G \mid 0^n \overset{\text{def}}{=} -(0^n \mid G), \ \text{typically } G \geq 0$		2.5, 4.1

Let H be an infinitesimal with canonical form $H = 0 \mid J$:

$H^0 \overset{\text{def}}{=} -J$ $H^i \overset{\text{def}}{=} 0 \,\Big|\, (-H^0 - H^1 \cdots - H^{i-1})$ $H^{\to 0} \overset{\text{def}}{=} 0$ $H^{\to i} \overset{\text{def}}{=} H^{\to i-1} \,\Big|\, J$ $\quad = H^1 + H^2 + \cdots + H^i$	H ith (e.g., H second)	4.11

Figure E.8: *Summary of definitions of games*

Examples (see pp. 189–192 for others)		
	$\multimapinv y$	First plays should be attacks on minies and big gote moves. Better to attack a tiny, creating a $y + 2$ point gote move than to cash in on a $y + 2$ point gote.
or   $\begin{cases} \\ \\ \end{cases}$	or $\begin{cases} \{y \mid 0\} \\ \{0 \mid -y\} \end{cases}$	
	$\multimapinv x$	Attacks on minies whose threats are smallest are worth least.
or   $\begin{cases} \\ \\ \end{cases}$	or $\begin{cases} \{x \mid 0\} \\ \{0 \mid -x\} \end{cases}$	
	$\multimapinv_x \mid 0$	Next, attack long corridors ending in bigger than 2 points gote moves. (Shorter corridors are attacked first if the gote moves at the end are equal.)
	$\multimapinv_x \mid 0^2$	
	$\multimapinv_x \mid 0^3$	
$\vdots$	$\vdots$	
	$\multimapinv_y \mid 0$	If two longer corridors are around, **attack the corridor with fewer stones at the end first!** (Approach the two stone group before the three stone group.)
	$\multimapinv_y \mid 0^2$	
	$\multimapinv_y \mid 0^3$	
$\vdots$	$\vdots$	
Continued on following page		

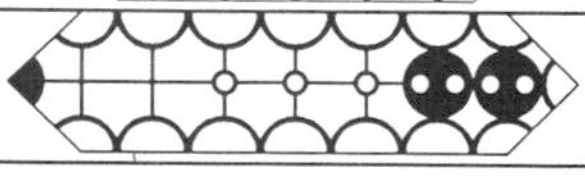

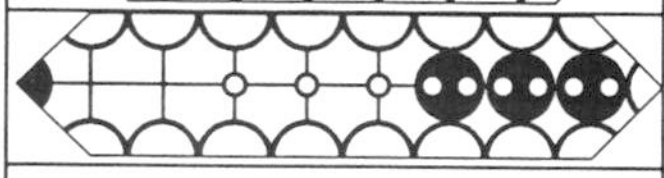

Figure E.9: *Which corridors Black should attack first. Here, x and y are numbers with $y > x > 0$. In the examples, $x = 2$ and $y = 4$. Remember to ignore any pair of chilled games summing to zero. Positions higher in the table are more urgent for Black.*

Continued from prior page		
Examples (see pp. 189–192 for others)		
or $\left\{\vphantom{\begin{matrix}a\\b\\c\end{matrix}}\right.$ $\vdots$	or $\left\{\vphantom{\begin{matrix}a\\b\\c\end{matrix}}\right.$ $\downarrow$ $\Downarrow *$ $\vdots$	Next attack corridors ending in one stone. Each such move is worth the same amount.
or ?	$*$(maybe)	But the move which connects the stone is worth less! Play a move on $*$ if an odd number of them are around.
or $\left\{\vphantom{\begin{matrix}a\\b\end{matrix}}\right.$	$\uparrow$ or $\Uparrow *$ or $\ldots$ or $0^n|\!+_x$	Block *any* corridor ending in at least a two point gote.
$\vdots$	$\vdots$	
or $\left\{\vphantom{\begin{matrix}a\\b\end{matrix}}\right.$	or $\left\{\vphantom{\begin{matrix}a\\b\end{matrix}}\right.$ $\frac{1}{4}$ $-\frac{1}{4}$	Numbers are low priority. Play the number with larger denominator first.
or $\left\{\vphantom{\begin{matrix}a\\b\end{matrix}}\right.$	or $\left\{\vphantom{\begin{matrix}a\\b\end{matrix}}\right.$ $\frac{1}{2}$ $-\frac{1}{2}$	
or	$ⓚ$ or $\overline{ⓚ}$	In Japanese rules, one-point ko's come next.
In non-Japanese scoring, these kos are worth more than one point, but the kofights should be started earlier only if the number of these kos is not divisible by three, and if the net number of kothreats exceeds half the number of dame. (See Section 5.3.)		
	0	Dame are played last.

Blocked corridors

The following table (Figure E.10) summarizes blocked corridor invasions by connected or unconnected invading groups. A *general blocked corridor* looks like:

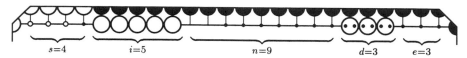

$$s = \text{socket depth}$$

s = socket depth
i = invading group's size
n = number of empty nodes adjacent to the invading group
d = dead group's size
e = empty nodes beyond the dead group

The examples in the table should clarify the following marking conventions.

- If there is a socket (i.e., $s, i \geq 1$), one node (the socket itself) is marked black, and $s - 1$ nodes are marked white. If, however, there is no invading group, all but two of the s-nodes are marked white.

- The invading group (represented by i) is unmarked.

- Of the empty nodes adjacent to the invading group, $n - 2$ are marked black.

- Each of the dead group's stones is marked twice, except if $n = 1$, in which case the first d-stone is marked only once.

- The e-nodes are marked black.

s	i	n	d	e	Example	Value
0	0	≥ 1	0	0		2^{1-n}
0	0	≥ 1	1	0	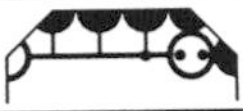	$*+(n-1)\cdot\uparrow*$
0	0	≥ 1	1	≥ 1		$0^n\,|-x$
0	0	≥ 1	≥ 2	≥ 0		$0^n\,|-x$
0	0	≥ 3	≥ 2	≥ 0	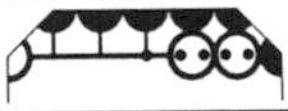	$0^{n-2}|+_x$
1	0	2	≥ 1	0	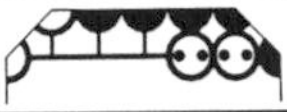	$\frac{1}{2}$
1	0	2	≥ 1	≥ 1	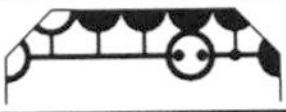	$0\|\textcircled{k}\,|-x$
1	0	≥ 3	≥ 0	≥ 0	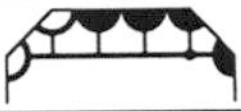	$0\|\textcircled{k}\,|\,1$
1	1	0	0	0		$\textcircled{k}-1$
1	1	1	0	0		0
1	1	≥ 2	≥ 0	≥ 0	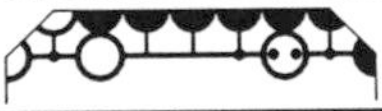	$\{\textcircled{k}-1\}\,|\,0$
≥ 1	≥ 2	≥ 2	≥ 0	≥ 0		—
≥ 2	0	≥ 2	≥ 0	≥ 0	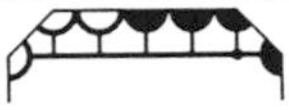	$*$
2	1	0	0	0		$*$
2	1	≥ 2	≥ 0	≥ 0	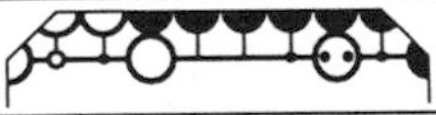	$\downarrow$
≥ 3	1	0	0	0	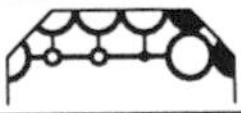	$0\,|\,1-\textcircled{k}\|0$
≥ 3	≥ 1	≥ 2	≥ 0	≥ 0		—

In the table, x is a positive number: $x=2d-2^{1-e}$.
"—" is given a liberal interpretation to include $0\,|-\textcircled{k}\|0\,\|\|\,0\approx-\frac{1}{6}$.

Figure E.10: *Summary of blocked corridor invasions by connected or unconnected invading groups*

Unblocked corridors

Next, we'll look at *general unblocked corridor invasions* in which the corridor is approached from both ends. These will look like two of the blocked corridor invasions placed back to back:

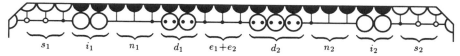

(To keep this diagram legible, each parameter is either 2 or 3.)

We'll consider several interesting cases.

If $s_1 > 0$ and $s_2 > 0$:

> **If, in addition, $n_1, n_2 \geq 2$:**
>
>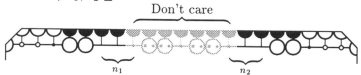
>
> Then the dead stones have no effect, and the unblocked corridor effectively decomposes into two blocked corridors. The value can be calculated by assuming the corridors were both blocked with $n_1 = n_2 = 2$. White's invasion on either end reverses through Black's block to White filling the socket:
>
>

If there are no dead stones and $n_1 + n_2 \geq 3$:

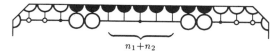

In like fashion, the corridor effectively decomposes, and its value is independent of $n_1 + n_2$.

If there are no dead stones and $n_1 + n_2 = 2$:

The chilled value is 0.

If $s_1 = 0$ and $s_2 = 0$ and $n_1, n_2 \geq 2$: Orient the corridor so that $d_1 \leq d_2$:

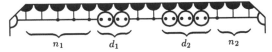

The value is of the form $0^{n_1+n_2-4}|\boldsymbol{+}$. White's canonical move is at the leftmost end, approaching the smaller group of dead stones first:

His penultimate rescue move reverses through Black's block to a White approach from the opposite end:

If $s_1 = 0$, $s_2 > 0$ and $n_2 \geq 2$: Label the three nodes adjacent to the invading groups as below:

Frequently this position can be simplified to a previously analyzed position by placing a black stone at a and a white stone at b:

For this simplification to yield an equivalent position, two conditions must hold:

1. Black's move at a must reverse through White's b.

2. White's move at c must reverse through Black's a.

We conjecture that these two conditions may be equivalent. However, the conditions under which these reversals hold are surprisingly subtle. If there are many dead or nearly dead strings, $d_1, d_2, d_3, \ldots$, the reversibility can depend on nearly all of their sizes and numbers of spaces between them.

For instance, let $s_2 = 2$. After the simplification, let the remaining **unchilled** game be denoted as

$$x_0 \; \Big|\Big|\Big|\Big| \; x_1 \; \Big|\Big|\Big| \; x_2 \; \Big|\Big| \; \cdots x_l \mid x_l$$

The condition $n_2 \geq 2$ ensures that the rightmost pair of values are equal. Using the fact that $0 = *\mid* = *\|0 \mid 0$, we may rewrite the value of the chilled game as

$$x_0 \; \Big|\Big|\Big|\Big|\Big| \; x_1 \; \Big|\Big|\Big|\Big| \; \cdots x_n \; \Big|\Big|\Big| \; x_{l+1}* \; \Big|\Big| \; x_{l+2}* \mid x_{l+3},$$

where $x_n = x_{l+1} = x_{l+2} = x_{l+3}$

To decide whether or not the reversals which yield the simplification are valid, let M denote the largest integer for which all m in the range $0 \leq m < M$, we have:

$$x_{m+2} \geq x_m - 2i_2 \geq x_{m+3} \qquad (E.1)$$

Since $i_2 > 0$, such an M exists, and in fact $M \leq l$. Since m cannot be enlarged, either

$$x_{m+2} < x_m - 2i_2$$

or

$$x_m - 2i_2 < x_{m+3}$$

In the latter case, both reversals are valid and the simplification holds. But in the former case, neither reversal is valid and the simplification fails.

Two examples, G_1 and G_2, should help to clarify the notation. H_1 and H_2 are the simplified games and, as we'll see, $G_1 \not\geq H_1$ but $G_2 = H_2$.

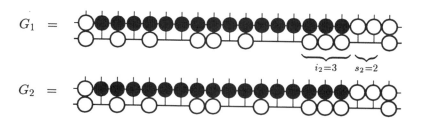

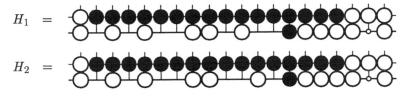

$$H_1 =$$

$$H_2 =$$

Ignoring the white (marked) point in the lower right corner of both simplifications, we have:

$$H_1 = 16 \;||||||||\; 13 \;|||||||\; 10 \;|||||\; 9 \;|||||\; 4 \;||||\; 1 \;|||\; 0 \;||\; * \;|\; 0 \cdots$$

$$H_2 = 16 \;||||||||\; 13 \;|||||||\; 10 \;|||||\; 9 \;|||||\; 4 \;||||\; 3 \;|||\; 0 \;||\; * \;|\; 0 \cdots$$

		$m \to$	0	1	2	3	4	5	6	
H_1	x_{m+3}		9	4	1	0	*	0	*	$\cdots$
	$x_m - 2i_2$		10	7	4	3	-2	-5	-6	$\cdots$
	x_{m+2}		10	9	4	$\triangleright\,1$	0	*	0	$\cdots$
H_2	x_{m+3}		9	4	3	0	$\triangleright\,*$	0	*	$\cdots$
	$x_m - 2i_2$		10	7	4	3	-2	-3	-6	$\cdots$
	x_{m+2}		10	9	4	3	0	*	0	$\cdots$

Condition (E.1) says that within each column the entries should not decrease. For each of H_1 and H_2, the first violation of this inequality is marked by a $\triangleright$. So our assertion implies that $G_1 \not\preceq H_1$ but $G_2 = H_2$.

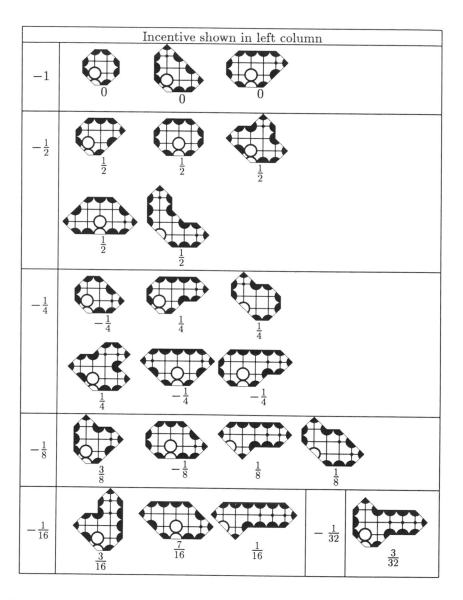

Figure E.11: *Rooms which chill to numbers. All ≤ 7 node rooms are sorted by shape in Section 4.5.*

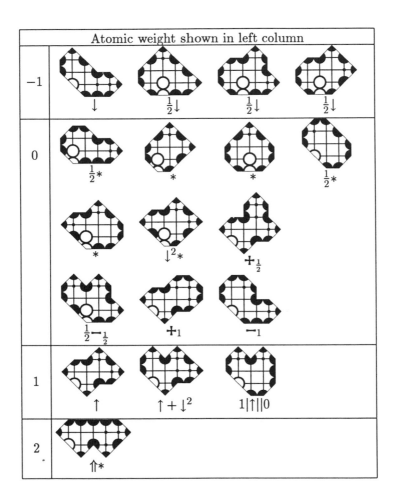

Figure E.12: *Rooms which chill to values infinitesimally close to numbers*

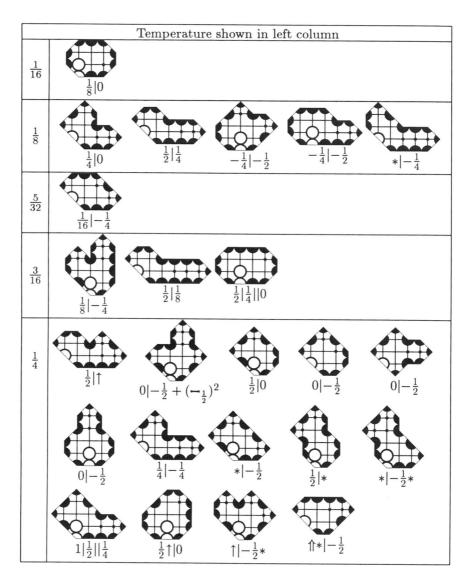

Figure E.13: *Hot rooms*

205

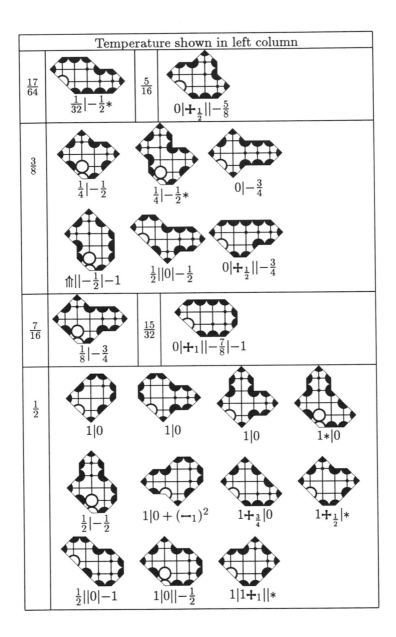

Figure E.13: (continued from prior page)

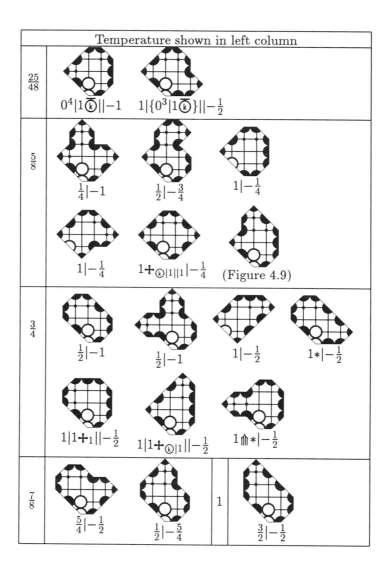

Figure E.13: (continued from prior page)

Appendix F

Glossary

The source of each word is identified by one of the following:

- [Games] : A word from combinatorial game theory.

- [Go] : A word in Go vocabulary, compatible with official rules in China, Taiwan, New Zealand, and North America.

- [Math] : A mathematical term.

- [Book] : A term defined in this book.

- [Rules **J**] : A term in the Official Japanese Nihon Ki'in rules [Boz92, pp. 230–242]. These tend to be more intricate definitions, which differ from [Go] only in exotic terminal positions in which Japanese rules may yield slightly different scores than anywhere else. (See Section B.5.7.)

ancient Go: [Book] A ruleset played in China circa 1000 A.D. It differed from modern Go primarily in deducting two points from each player's score for each of his separate live groups. (Section B.2)

atari: "ah-tah-ree" [Go] If the number of liberties of an opponent's string is one, the group is said to be in atari. Often, in beginners' games, a player who reduces the number of liberties of an opposing string to one says "atari" to warn of imminent capture. The word is analogous to "check" in chess.

atomic weight: (or uppitiness) [Games] About how many ↑'s something is worth. An atomic weight of one implies an infinitesimal is a win for

Left moving first. An atomic weight of two or more implies it is greater than zero, and therefore a win for Left even if Right moves first. For a formal definition, see [BCG82, p. 223] or [Con76, p. 219].

b-value: [Book] Short for the chilled and normalized value of a blocked corridor. (Section 4.3)

Belle: [Book] Belle Black is the character playing the black stones against Wright White. (See also Left:, Right:)

big eye: [Go] Black's big eye is a region enclosed by Black. If White begins play within the big eye, he can ensure that Black gets only one eye in this region. If Black is allowed to play (possibly several times consecutively) within this region, she can make two eyes. (Section B.5.3)

canonical form: [Games] The unique minimal sized game tree equal to a given game, obtainable by repeatedly removing dominated options and reversing reversals. (Section 3.5.2)

capture: [Go] A play which reduces the number of liberties of an opposing string to zero *captures* the group. The string is removed from the board and held as prisoners. (See also liberty:, string:, group: and Section B.2.1.)

chill: [Games]

$$f(G) \overset{\text{def}}{=} \begin{cases} n & \text{if } G \text{ is of the form } n \text{ or } n* \\ \left\{ f\left(G^L\right) - 1 \ \middle| \ f\left(G^R\right) + 1 \right\} & \text{otherwise} \end{cases}$$

and for elementary Go positions is proven equivalent to G cooled by 1 in Theorem 1.6. (Sections 2.2 and 3.6)

cold: [Games] A game is cold if it equals a number. (Section 3.5.3)

competent player: [Book] A player who will play a position flawlessly. The positions with which we challenge the competent player are, in our opinion, within the range of most experienced Go players ranked 10 kyu or higher. (See also guru:)

confirmation phase: [Rules J] An analysis phase applied after all play in the game is completed. It is used to confirm assertions about life, death, eye points, dame, seki, and territory as may be needed to calculate the official Japanese score and to determine the result. The confirmation phase uses special Ko-playing rules. (Section B.4.6)

cooling: [Games] See Section 3.5.4. (See also unheating:)

dame: "dah-meh"

1. [Go] A move on the border of opposing live groups which neither earns nor loses points by Japanese and American scoring rules. (See center row of Figure B.7.)

2. [Rules J] "Empty points that are not eye points are called dame." (Section B.5.7)

dan: "dahn" [Go] A rank of a strong Go player. Amateur dan ranks go from 1-dan to 7-dan, and professional are on a separate scale running from 1-dan to 9-dan. Ranks below 1-dan amateur are termed *kyu* ranks, with n-kyu being the number of ranks below 1-dan. So, 1-kyu is the rank immediately preceding 1-dan.

dead stone:

1. [Go] A stone which is destined to be captured if play continues into a competently played encore.

2. [Rules J] "Stones which are not alive are said to be dead." (See life: and Section B.5.7.)

defective shape:

1. [Book] A shape which allows one player to gain by playing onto the board from a position which superficially looks like a TGAP. (Section B.5.3)

2. [Go] A shape which is easily attacked.

dominant move: [Games] An option which dominates all others is said to be *the dominant move.* If there is a set of several moves, none of which is dominated by any others, then this is the *set of dominant moves.*

dominated option: [Games] Left option G^{L_1} is dominated if there exists a G^{L_2} s.t. $G^{L_2} \geq G^{L_1}$. G^{R_1} is dominated by a $G^{R_2} \leq G^{R_1}$. (Section 3.5.2)

dyadic rational: [Math] Any number of the form $\frac{i}{2^j}$ for i and j integers.

elementary: [Book] A Go game without kos or odd sekis. (Section 3.6)

encore: [Book] A terminal, *score-counting* stage of play that is traditionally omitted from actual games. (Section B.4.2)

equals: [Games] $G = H$ if the second player wins the difference game $G - H$. (Section 3.5.1)

even: (See parity:)

eye point:

1. [Go] See pair of eyes: and private eye:

2. [Rules **J**] "Empty points that are surrounded by the live stones of just one player are called eye points." (Section B.5.7)

false eye: [Go] See Figure B.8(b).

follower: [Games] A game's followers are the positions arrived at after a single move by either player. (Compare with position:)

game: [Games] Either a specific (local) game position, or the rules for a game. (Section 3.5.1)

game tree: [Games] A representation (often pictorial) showing the possible sequences of moves from a particular game position. Since one player may make consecutive moves on a local position, we need to distinguish the players' moves. Left's (i.e., Black's) moves are shown by arrows pointing toward the left on the page, and Right's moves point rightward. See, for example, Figure 2.11 on page 23.

gote: "go-teh" [Go] A locally non-forcing move.

greater-equals: [Games] $G \geq H$ if Left wins the difference game $G - H$ moving second. (Section 3.5.1)

greedy: [Book] A Go-playing strategy which refuses to pass if there exists any move on the board which appears to immediately improve one's score in a precisely defined sense. (Section B.2.4)

group:

1. [Go] Group is the unit of life. However, in the Go literature, the meaning of *group* is imprecise. Loosely, a group is a collection of strings that will live or die together, even if one or more strings of the group may be sacrificed. (See also string:.)

2. [Book] In this book we've limited ourselves precisely to collections of strings which share liberties in a terminal greedy ancient position.

3. [Math] A set which (a) is closed under an associative binary operation, (b) contains an identity, and (c) has an inverse for every element.

guru: [Book] A perfect Go player. When faced with relatively unchallenging positions, we may call him merely *competent*. Neither competent players nor gurus ever play suboptimally.

handicap: [Go] In a Go game between players of differing ranks, the weaker player is often given an advantage of several moves to start the game. These are called handicap stones and, according to some rulesets, must be placed at standard positions.

heating: [Games] G heated by t is denoted by $\int^t G$. If the canonical form of $G = G^L | G^R$ and if t is a number, then

$$\int^t G \overset{\text{def}}{=} \begin{cases} G & \text{if } G \text{ is a number} \\ \left\{ t + \int^t G^L \ \middle| \ -t + \int^t G^R \right\} & \text{otherwise} \end{cases}$$

If t is negative, this definition is said to *unheat* G by $-t$.

hot: [Games] A game is hot if its left stop exceeds its right stop. (Section 3.5.3)

hung: [Book] A hung outcome to a game ends without a result due to unusual positions such as triple-ko. For more details see Section A.5.

immortal: [Book] A stone is immortal if it cannot be captured for the rest of the game. This may be a consequence of the ruleset or simply an assumption for analysis.

incentive: [Games] The set $\Delta^L \{G\} = G^L - G$ are the left incentives of G. These indicate the value of a move to Left. $\Delta^R \{G\} = G - G^R$ are Right's incentives and $\Delta \{G\} = \Delta^L \{G\} \cup \Delta^R \{G\}$ are the incentives of a game. (Section 3.5.5)

incomparable: [Games] $G \not\parallel H$ (read "G is incomparable to H") if G is neither greater than, less than, nor equal to H. (Sections 2.4 and 3.5.1)

independently live group: [Go] Precisely, in a TGAP, an independently live group is a set of strings whose reduced topological graph is a single branch.

infinitesimal: [Games] A game less than all positive numbers and greater than all negative numbers. Alternatively, a game whose Left and Right stops are both zero. G is infinitesimally close to H or H-ish if $G - H$ is an infinitesimal. (Sections 2.3 and 3.5.3)

ish: [Games] Infinitesimally SHifted. (See infinitesimal:)

ko: "kō" [Go] A position in Go which repeats by the capture and recapture of a single stone. In the simplest one-point ko, there are no other side effects dependent on who gets to capture. (Sections B.2.1 and 4.5)

komi: "kō-mee"

1. [Go] A number of points given to White to compensate for moving second. Typical values of komi are $5\frac{1}{2}$, $6\frac{1}{2}$ and $7\frac{1}{2}$ and vary from tournament to tournament. In clubs, komi is sometimes used as a method of handicap.

2. [Book] In combinatorial game theory, negative scores are defined to be good for White, so typical values of komi are $-5\frac{1}{2}$, $-6\frac{1}{2}$ and $-7\frac{1}{2}$.

kyu: (See dan:)

Left: [Games] One of the two players of a game in combinatorial game theory. Positive games are good for Left and Left plays the Black stones in Go. (Section 3.5.1)

less-equals: [Games] $G \leq H$ if Right wins the difference game $G - H$ moving second. (Section 3.5.1)

liberty: [Go] An empty intersection adjacent to a string. (See also string:, group:, capture:; Section B.2.1.)

life: 1. [Go] A group of stones is said to make life if it cannot be captured even if the opponent plays next and players alternate play thereafter.

2. [Rules **J**] "Stones are said to be alive if they cannot be captured by the opponent [in the confirmation phase of play], or if capturing them would enable a new stone to be played that the opponent could not capture." (Section B.5.7)

living eyespace: [Go] Black's living eyespace is a region surrounded by Black, within which a competent Black player can make two eyes even if White plays first.

mapping: [Math] A function from one set (called the domain) into another set (called the range). To each element in the domain, the mapping assigns precisely one element in the range.

markings: [Book] Small black and white dots placed on a board to normalize the score. (Section 4.1)

miai: [Go] "mee-eye" Two moves which are interchangeable. Whichever move one player makes, his opponent can make the other.

negative: [Games] A win for White (in Go) or Right (in combinatorial game theory) no matter who moves first. (Section 3.5.1)

Norton multiply: [Games] G Norton multiplied by U is defined in terms of the incentives of U [BCG82, p. 246]:

$$G \cdot U \stackrel{\text{def}}{=} \begin{cases} \overbrace{U + U + \cdots + U}^{G \text{ times}} & \text{if } G \text{ is an integer} \\ \left\{ G^L \cdot U + (U + \Delta\{U\}) \;\middle|\; G^R \cdot U - (U + \Delta\{U\}) \right\} \end{cases}$$

number: [Games] A game whose value is an integer (such as 3 or -17) or a dyadic rational ($2\frac{1}{2}$ or $-2\frac{2}{8}$). (Section 3.5.3)

odd: (See parity:)

odd seki: [Book] A seki whose parity is odd, i.e., with an odd number of empty intersections (not to be confused with bizarre positions such as Figure B.24, which are declared to be seki in ruleset **J**, but may not be considered to be seki in any other country).

one-to-one mapping: [Math] A mapping such that at most one element in the domain is mapped onto each element in the range. Cooling is a one-to-one mapping if its domain is taken as the set of even elementary Go positions in canonical form. (Section 3.6)

overheating: [Games] G overheated from s to t is defined in [Ber88] and is given by

$$\int_s^t G \stackrel{\text{def}}{=} \begin{cases} \overbrace{s+s+\cdots+s}^{G \text{ times}} & \text{if } G \text{ is an integer} \\ \left\{ t+\int_s^t G^L \mid -t+\int_s^t G^R \right\} & \text{otherwise} \end{cases}$$

pair of eyes: [Go] In a TGAP, the two nodes at the ends of the topological graph corresponding to an *independently alive* group.

parity: [Book] A Go position is *even* (or *odd*) if the number of empty intersections plus the number of prisoners captured is even (or odd). We also say the Go position's *parity* is even (or odd). (Section 3.6)

position: [Games] A game's positions are the set of games which can be arrived at by legal play by either or both players. As usual in combinatorial game theory, the same player may make several consecutive moves. (Compare with follower:)

positive: [Games] A positive game is a win for Black (in Go) or Left (in combinatorial game theory) no matter who moves first. (Section 3.5.1)

private eye: [Book] In a TGAP, the node at either end of a linear bichromatic topological graph.

purported incentives: [Book] Let G be a game, and say we conjecture that game $f(G) = G$. The purported incentives, denoted $\hat{\Delta}$, $\hat{\Delta}^R$ and $\hat{\Delta}^L$, or when chilled $\hat{\delta}$, $\hat{\delta}^R$ and $\hat{\delta}^L$, are the incentives of $f(G)$. They are used in a technique to prove $G = f(G)$. See Section 4.6.

rational: [Math] Any number of the form $\frac{p}{q}$ for p and q integers.

reversible: [Games] Left's move to G^L reverses through a Right response to G^{LRL} if $G^{LR} \leq G$. Left's move to G^L is said to be a *reversible move*. Similarly a $G^{RL} \geq G$ indicates a reversal. (Section 3.5.2)

Right: [Games] One of the two players of a game in combinatorial game theory. Negative games are good for Right, and Right plays the white stones in Go. (Section 3.5.1)

ruleset: [Book] Rules of Go vary slightly in differing regions of the world, but basic principles of play remain the same everywhere. A ruleset refers to the specific set of rules such as those analyzed in Appendices A and B. Abbreviations used in the text are:

A	ancient	**M**	All Mathematical dialects
C	modern China	**U=MU**	Universalist
J	Japan	**MJ**	Mathematized Japanese
NA	North American	**MC**	Mathematized Chinese
Ing	Ing's 1986 rules	**MNA**	Mathematized North
T	Taiwan		American
NZ	New Zealand		

seki: "seh-kee"

1. [Go] Situation in which bordering groups of black and white stones share liberty(s), yet the player who tries to initiate capture by playing the liberty(s) loses stones. A stalemate between two or more groups. (A precise definition appears in Section B.2.8.)

2. [Rules J] "Eye points surrounded by stones that are alive but not in seki are called territory." (Appendix B, especially Figures B.15 and B.24)

semeai: [Go] "seh-meh-eye" A capturing race between two opposing groups which (1) are both surrounded, (2) share liberties or a boundary, and (3) cannot make two eyes. Depending on the position, one player may capture the other, or the position may resolve to a seki.

sente: [Go] "sen-teh" A locally forcing move.

simplest number: [Games] The simplest number in an interval is the integer of lowest magnitude if one exists, or the dyadic rational of smallest denominator, otherwise. (Section 3.5.3)

snapback: [Go] A sequence of two consecutive moves, one by each player. The first move captures a single stone. The second move places a new stone onto the node just vacated by the captured stones under conditions which then capture more than one of the opponent's stones. (Section B.2.1)

socket: [Book] A play (eventually) required to connect a group to life; the node where this play must be made. (Section 4.7)

stop: [Games] A stop, or stopping position, is any game which is a number. The left (right) stop is the stop reached when left (right) moves first, with moves alternating and both players playing optimally. (Section 3.5.3)

string: [Go] A set of stones of one color connected to one another by the lines of the grid of a Go board. A string is the unit of capture. (See also liberty:, capture:, group:; Section B.2.1.)

tedomari: [Go] "teh-doh-mah-ree" The last move of the game, but often used to refer to the last big move or last good move, especially in the opening game and the endgame.

tepid: [Games] A game is tepid if it equals a number plus a non-zero infinitesimal. (Section 3.5.3)

terminal greedy ancient position: [Book] (abbreviated TGAP) A position in which for both players (a) no legal move captures any opposing stone(s), and (b) each legal move has a response which captures the string containing the just-played stone. (Sections B.2.4 and B.2.5)

terminal position: [Book] A position in which two opposing gurus would both choose to pass.

TGAP: [Book] terminal greedy ancient position.

topological configuration: See terminal greedy ancient position: and Sections B.2.5 and B.2.6

unheating: [Games] If t is a positive number, then G unheated by t is defined as G heated by the negative number $-t$ and is denoted $\int^{-t} G$.

uppitiness: [Games] See atomic weight:

warming: [Games] A warming operator is any overheating operator, $\int_s^t$, for s and t 1-ish (Section 3.6). In this book, the only warming operator used is $\int_{1*}^1$, and is equivalent to the Norton multiply by 1*. More precisely,

$$\int G = \begin{cases} G & \text{if } G \text{ is an even integer} \\ G + * & \text{if } G \text{ is an odd integer} \\ \left\{ 1 + \int G^L \mid -1 + \int G^R \right\} & \text{otherwise} \end{cases}$$

Wright: [Book] Wright White is the character playing the white stones against Belle Black. (See also Left:, Right:)

worthwhile: [Games] A worthwhile move is sufficient to win in any context in which a win is possible. In the mathematical game $G = \left\{ 0, \frac{1}{4} \mid 1 \right\}$, Left's move to 0 is worthwhile even though it is dominated by $\frac{1}{4}$.

Bibliography

[BCG82] Elwyn R. Berlekamp, John H. Conway, and Richard K. Guy. *Winning Ways*. Academic Press, New York, 1982.

[Ber88] Elwyn R. Berlekamp. Blockbusting and domineering. *Journal of Combinatorial Theory*, 49(1):67–116, September 1988.

[Ber91] Elwyn Berlekamp. Introductory overview of mathematical Go endgames. In *Proceedings of Symposia in Applied Mathematics: Volume 43, Combinatorial Games*, pages 73–100. American Mathematical Society, 1991.

[Boz92] Richard Bozulich. *The Go Player's ALMANAC*. Ishi Press, San Jose, Tokyo, and London, 1992.

[Che] Raymond Chen. Ishi Press International, 76 Bonaventura Drive, San Jose, CA 95134. Software which plays a collection of one-point Go endgame problems.

[Con76] John H. Conway. *On Numbers and Games*. Academic Press, London/New York, 1976.

[Edi] David Erbach (Journal Editor). Computer Go. 71 Brixford Crescent, Winnipeg, Manitoba R2N1E1, Canada.

[Fai90a] John Fairbairn. Go on the roof of the world. *Go World* (58):10–15, Winter 1990.

[Fai90b] John Fairbairn. Monster Go. *Go World* (60):46–48, Summer 1990.

[Fai92] John Fairbairn. Liu Zhongfu's Go secrets. *Go World* (67):61–64, Spring 1992.

[Gar74] Martin Gardner. Mathematical games: Cram, crosscram and quadraphage: new games having elusive winning strategies. *Scientific American*, 230(2):106–108, February 1974.

[GS56] Richard K. Guy and Cedric A. B. Smith. The *G*-values of various games. *Proceedings of the Cambridge Philosophical Society*, 52:514–526, 1956.

[Guy89] Richard K. Guy. *Fair Game: How to play impartial combinatorial games*. COMAP, Inc., 60 Lowell Street, Arlington, MA 02174, 1989.

[Han59] Olaf Hanner. Mean play on sums of positional games. *Pacific Journal of Mathematics*, 9:81–89, 1959.

[Har88] Haruyama Isamu 9-dan. Strange and wonderful shapes (triple ko, eternal life). *Go World* (50):16–22, Winter 1987-88.

[Hig91] Bob High. Mathematical Go. *Computer Go* (15):14–24, Fall/Winter 1990-91.

[Ike92] Ikeda Toshio. *On the Rules of Go*. Fujitsu, Ltd., Tokyo, Japan, 1992.

[Ing86] Ing Chang-ki. *The SST Laws of Wei-chi'i*. Ing Wei-chi'i Educational Foundation, 4th Floor, No 35 Kuagn Fu S. Road, Taipei, Taiwan, 1986.

[Ish] Ishi Press International. 76 Bonaventura Drive, San Jose, CA 95134. (408)944-9900.

[Kie90] Anders Kierulf. *Smart Game Board: a Workbench for Game-Playing Programs, with Go and Othello as Case Studies*. PhD thesis, Swiss Federal Institute of Technology (ETH) Zürich, 1990. Nr. 9135.

[Knu74] Donald E. Knuth. *Surreal Numbers*. Addison-Wesley, Reading, MA, 1974.

[KOM86] Kudo Norio 9-dan, O Meien 9-dan, and Murakami Akira. Why not eliminate the irrational in Go? *Go World* (45):60–64, Autumn 1986.

[Lai90]   Roy Laird. More about rules than you really want to know. *American Go Journal*, 24(2):9–13, Fall 1990.

[Lan]   Howard Landman. Personal communication.

[LS80]   David Lichtenstein and Michael Sipser. Go is polynomial-space hard. *Journal of the Association for Computing Machinery*, 27(2):393–401, April 1980.

[Mil53]   John Milnor. Sums of positional games. In Kuhn & Tucker, editor, *Contributions to the Theory of Games*, pages 291–301. Princeton, 1953.

[Moe]   David Moews. To appear in Ph.D. thesis from University of California, Berkeley, 1993.

[Mou90]   David Moulton, November 1990. Personal communication, including oral report given in Professor Berlekamp's Math 275 class at University of California, Berkeley.

[Nak84]   Nakayama Noriyuki 6-dan. *The Treasure Chest Enigma*, page 122. Ishi Press, 1984. Problem 20: The Bait Swallows the Fish by Honinbo Dosaku [1645-1702].

[Nak91]   Nakayama Noriyuki 6-dan. Strange and wonderful shapes. *Go World* (64):61–64, Summer 1991.

[OK91]   Ohira Shuzo 9-dan (Black) and Komatsu Hideki (then) 7-dan (White). *Igo Nenkan,* p. 190, yearly supplement to the monthly magazine, *Kido,* pubished by Nihon Ki'in, Japan, 1991. Game played on January 18, 1990 at the Nihon Ki'in.

[Pem84]   Robin Pemantle. Undergraduate honors thesis in the Department of Mathematics at the University of California at Berkeley, 1984.

[Rob83]   J. M. Robson. The complexity of Go. In *Proc. IFIP (International Federation of Information Processing)*, pages 413–417, 1983.

[Sak81]   Sakai Takeshi 9-dan. *A Selection of Cho's Good Games*, volume 2. Nihon Ki'in, Japan, 1981. In Japanese. ISBN = 4-8182-0205-3.

[Sho93]   Peter Shotwell. Go in the snow. *Go World* (69), Autumn 1993.

[Tak90]   Takeshiro Yoshikawa. The most primitive Go rule. *Computer Go* (13):6–7, Winter 1989-90.

[Wol]   David Wolfe. Games program available via anonymous ftp from bsdserver.ucsf.edu in the file Go/comp/theory.sh.Z, or send e-mail to wolfe@cs.Berkeley.EDU.

[Wol91]   David Wolfe. *Mathematics of Go: Chilling Corridors.* PhD thesis, Division of Computer Science, University of California at Berkeley, 1991.

[Wol93]   David Wolfe. Snakes in domineering games. *Theoretical Computer Science*, 119(2):323–329, October 25 1993.

[Yed85]   Laura Yedwab. On playing well in a sum of games. Master's thesis, M.I.T., August 1985. MIT/LCS/TR-348.

Index

About the Authors

Elwyn Berlekamp has been Professor of mathematics and of electrical engineering-computer science at UC Berkeley since 1971. He has founded or co-founded several successful companies, including Cyclotomics, now known as Kodak Berkeley Research. He has also developed and refined some techniques which are used to manage investment portfolios profitably. He has 12 patented inventions and over 75 publications, including two original books: *Algebraic Coding Theory,* McGraw-Hill 1968 and Aegean Park Press, 1984; and *Winning Ways,* coauthored by Conway and Guy, Academic Press 1982. Since grade school, mathematical games have been among his foremost interests.

David Wolfe is a lecturer in theoretical computer science at UC Berkeley. He is a former president of the Berkeley Go Club, and he has also been active in teaching Go at elementary schools in the Bay Area. A significant portion of this book is based on his 1991 PhD thesis. His other research interests include combinatorics, stochastic process, randomness and computation.

James Davies, who wrote the foreword, is the world's best known and most prolific writer of English-language Go books, several of which are co-authored by well-known professional Japanese Go players. He is also a professional Japanese/English translator and interpreter, a 6-dan Go player, and one of the world's foremost authorities on the rules of Go. His works include 11 titles published by the Ishi Press.